SMALL ISLANDS IN PERIL?

Island Size and Island Lives in Melanesia

SMALL ISLANDS IN PERIL?

Island Size and Island Lives in Melanesia

EDITED BY COLIN FILER

Australian
National
University

ANU PRESS

ASIA-PACIFIC ENVIRONMENT MONOGRAPH 18

Australian
National
University

ANU PRESS

Published by ANU Press
The Australian National University
Canberra ACT 2600, Australia
Email: anupress@anu.edu.au

Available to download for free at press.anu.edu.au

ISBN (print): 9781760466534
ISBN (online): 9781760466541

WorldCat (print): 1439078687
WorldCat (online): 1439079004

DOI: 10.22459/SIP.2024

Cover design and layout by ANU Press
Cover photograph: Masahet Island by Nicholas Bainton

This book is published under the aegis of the Asia-Pacific Environment Monographs editorial board of ANU Press.

Contents

List of Tables

List of Figures

Contributors

Nicholas Bainton is an Associate Professor in the School of Regulation and Global Governance at The Australian National University.

Colin Filer is an Honorary Professor in the Crawford School of Public Policy at The Australian National University.

Simon Foale is an Associate Professor in the College of Arts, Society and Education at James Cook University.

Edvard Hviding is a Professor in the Department of Social Anthropology at the University of Bergen.

Jeff Kinch is a Research Fellow in the Pacific Livelihoods Group at Curtin University.

Nancy Lutkehaus is a Professor of Anthropology and Political Science in the Dornsife College of Letters, Arts and Sciences at the University of Southern California.

Martha Macintyre is a Principal Research Fellow in the School of Social and Political Sciences at the University of Melbourne.

1

Introduction

Colin Filer

A Personal Policy Odyssey

I first became interested in the problems of small islands under pressure when I was involved in the process of assessing the future impact of the Lihir gold mine in the 1980s. The Lihir group of islands in New Ireland Province consists of one medium-sized island, sometimes called Lihir but known locally as Niolam or Aniolam, and four very small islands, three of which are permanently inhabited. In 1980, two of these very small islands (Malie and Masahet) were already under pressure, while the third (Mahur) had not yet reached the critical level of population density. In order to refine the concept of 'pressure', my colleague, Richard Jackson, and I worked out how much land was required to maintain the existing system of food production on these islands, and how much would be required in future if the population were to grow at the rate of 2 per cent a year. With some other assumptions added to the mix, we figured that the people of Malie and Masahet:

> should already be suffering from a noticeable shortage of land, and ... these problems will become acute within the next two or three decades unless there is a radical transformation of the agricultural system, a substantial reduction in the birth rate, or an equally substantial increase in the rate of outmigration.
>
> (Filer and Jackson 1989: 38)

This was one part of a baseline scenario in which we tried to predict what would happen in the absence of the gold mine whose impact was being assessed. Since then, the gold mine has actually been developed, and all sorts of social, economic and environmental changes have taken place as a result (Macintyre and Foale 2004). So these three small islands have experienced new forms of pressure that set them apart from all the other very small island communities of Papua New Guinea (PNG), where there is no huge gold mine in the immediate vicinity.

In the second half of the 1990s, when I was working at the National Research Institute in Port Moresby, the problems being encountered on some of PNG's other small islands were drawn to my attention by a number of visiting anthropologists. Boang Island in New Ireland Province, M'buke Island in Manus Province and Brooker Island in Milne Bay Province are three examples that come to mind. These were essentially stories about population pressure on terrestrial and marine resources. At the same time, the national newspapers were already beginning to publish stories about the fears being expressed in PNG's coral atoll communities about the prospect of sea level rise as a result of global warming (Field 2000). Although the people of the Carteret (or Tulun) Islands have since captured a bigger share of international attention on this score, those of the Mortlock (or Takuu) Islands and the Tasman (or Nukumanu) Islands are in much the same predicament, and in one sense are worse off because their atolls are even more remote from terra firma. While the Carteret Islands are 'only' 100 km from Buka and Bougainville, the distance to the Mortlock Islands is 270 km, and the distance to the Tasman Islands is 550 km. However, if these two 'Polynesian' communities are the most remote of PNG's very small island communities, the Carteret Islands are under greater pressure or in greater peril because they hold the national record for small island population density—more than $1,224/km^2$ in 2000 (Bourke and Betitis 2003: 14).

Despite the publicity attached to the plight of these three island groups, it is important to remember that most of the very small islands under pressure in PNG are not coral atolls, and sea level rise is not the biggest problem that they face—at least not yet. In the case of the Carteret Islands, it is also worth recalling that the rate of population growth was already thought to be unsustainable in the colonial period, long before anyone heard of global warming, and the colonial administration already had plans to solve this problem by resettling part of the population on Buka or Bougainville. Narratives of climate change should therefore not distract us from asking what it is that turns a certain rate of population growth or level of population density into a perilous situation, how people have adapted to such situations in the past, and how their options have been changing through the colonial and post-colonial period.

In September 2000, an economist with whom I had previously collaborated in the implementation and evaluation of PNG's Biodiversity Conservation and Resource Management Program (BCRMP) drew my attention to a call for proposals to conduct 'sub-global assessments' as part of the Millennium Ecosystem Assessment (MA). A fellow anthropologist who had also been involved in the evaluation of the BCRMP warned me that the MA was likely to be a purely technocratic exercise that would not be amenable to our way of thinking. Foolishly perhaps, I ignored his advice and took this as an opportunity to think more deeply about the problems of small islands under pressure in PNG. In my initial expression of interest, I narrowed the definition of 'small islands' to those with a surface area of less than 10 km^2, but also suggested that these were representative of a much larger group of such entities spread across the Indo-Pacific region. This does not seem to have impressed the selection committee that was responsible for deciding which parts of the globe were most deserving of a sub-global ecosystem assessment (Filer 2009: 89). And matters might have rested if the economist had not found a way to incorporate my proposal into the design of the Milne Bay Community-Based Coastal and Marine Conservation Project (MBCP). It was his idea to change 'small islands under pressure' into 'small islands in peril' (SMIP). The second phrase was not only a better fit for the acronym but also resonated with the title of a report about the threat that climate change posed to the biodiversity of coral reef ecosystems across the whole of the Pacific Island region (Hoegh-Guldberg et al. 2000).

The MBCP was in effect the successor to the BCRMP, in the sense that it represented a second attempt by the Global Environment Facility to inject a large amount of money into PNG's biodiversity conservation business. The first attempt had come to a rather sticky end in 1998 when it was found that PNG's 'rainforest people' would only exhibit a preference for biodiversity conservation over resource development if the second option was not available (van Helden 2001, 2009; Filer 2004). The response to this discovery was to leave the rainforest people to their own devices, and instead fund the design of a marine conservation project in a coastal province that had relatively high levels of formal education and labour mobility, but where local communities were exerting unsustainable fishing pressure on coral reef ecosystems (Kinch 2001; van Helden 2004). Like the BCRMP before it,

3

the MBCP would be 'implemented' by the United Nations Development Programme (UNDP), but instead of being 'executed' by the PNG Government (through the Department of Environment and Conservation), it would instead be executed by a big international non-governmental organisation (NGO)—Conservation International—notionally working in partnership with the Milne Bay Provincial Government. As matters turned out, the MBCP was a more spectacular failure than the BCRMP. Five years of expensive design were followed by three years of expensive implementation and execution, and then the project was executed in another sense before it could proceed to its second phase (Baines et al. 2006; Balboa 2009, 2013).

From the point of view of the UNDP, the inclusion of our 'ecosystem assessment' as a separate component of the MBCP made sense for three reasons. The first was that the MA was itself a UN initiative, personally endorsed by Kofi Annan. The second was that it offered a way of investigating society–environment relationships in the target area that would go beyond the narrow focus of the executing agency on the establishment of a network of marine protected areas. The third was that it would be undertaken by social and natural scientists who were not themselves employed by the executing agency, and might therefore be able to make an independent contribution to understanding the social and environmental context and impact of the project.

From my point of view, simultaneous engagement with the MA and the MBCP proved to be a mixed blessing. The promise of funding from the UNDP made it possible to leverage approval (and some additional resources) from the MA to assemble a group of experts to start work on a sub-global assessment exercise while we waited for the promise to be kept. This was fortunate because the promise was not kept until the second half of 2004, but it was also problematic because all 'approved' sub-global assessments were required to report their findings before the end of that year. As it became apparent that our team would not be able to report the findings of new research in Milne Bay within the prescribed time frame, we were authorised to undertake a preliminary assessment of 'coastal, small island and coral reef ecosystems' across the whole of PNG, with a specific focus on a number of local areas (including parts of Milne Bay) where a substantial amount of relevant information was already available.

There is little doubt that we bit off more than we could chew by including all 'coastal' ecosystems within the scope of this assessment. The total length of the PNG coastline is in excess of 17,000 km. Even if the 'coastal zone' is defined quite narrowly, as the space that extends 10 km inland or offshore, it encompasses 10 per cent of the country's landmass, and more importantly is home to roughly one-third of the country's total population. As a result, we had now got into the business of assessing the consumption of ecosystem services by coastal communities containing about 2 million people instead of the 100,000 or 200,000 who might be living on SMIPs. This enlargement of scope might have been justified by a desire to make the results of the assessment seem more important to national policy makers, or to see whether 'small island ecosystems' were actually under any greater pressure than other 'coastal ecosystems'—such as those located in proximity to large coastal towns. But in practice, it was grounded in a lack of certainty about what constitutes an 'ecosystem' in the first place.

The MA conceptual framework treated 'coastal' and 'island' ecosystems as two out of ten broad categories of ecosystem, but allowed that any particular 'place on Earth' might belong to more than one of these ten classes (MA 2003: 54). Small island ecosystems could thus be thought of as a subset of coastal ecosystems. However, while a small island and the waters surrounding it may count as a single 'place on Earth', it is also a bundle of different ecosystems if ecosystems are defined as the biological communities that are associated with a certain type of physical environment. So one island ecosystem might include an area containing coconut orchards, another area containing mangrove swamps, a third area containing coral reefs, and so on. Each of these small areas would then count as part of a much larger area containing biological communities of the same type, and these would not necessarily be associated with small islands. If we then set out to assess the condition of one type of biological community within the boundaries of a single nation state on the basis of available scientific data, it might simply prove impossible to distinguish between those that are associated with small islands and those that are not. So it turned out that the conduct of an ecosystem assessment at a national, as well as a local, scale actually required us to enlarge the scope of the assessment, either to embrace all of PNG's inshore marine ecosystems, or else to embrace all of PNG's coastal ecosystems (Filer et al. 2004). It might have suited the needs of the MBCP if we had in fact concentrated on the first of these categories, and we added 'coral reefs' to the title of our sub-global assessment as a gesture in this direction, because this type of marine ecosystem was of central concern to the project. However, small island communities are not exclusively dependent on the services supplied by marine ecosystems; they also make fairly exhaustive and sometimes unsustainable use of terrestrial resources.

Most anthropologists are not used to bothering too much about the definition and classification of ecosystems, but they have some reason to be bothered by a vision of the world in which human agency finds its primary expression in the consumption of ecosystem services, scientific or economic reflection on such acts of consumption, and the production of decisions about how they should be managed. Yet that is precisely the sort of world that was envisaged in the MA conceptual framework. This is an elaborated form of the pressure–state–response model familiar to biologists and ecologists, and was meant to facilitate a productive dialogue between such people and environmental or ecological economists. It was not meant to appeal to anthropologists, let alone to represent the local knowledge of Melanesian villagers (Filer 2009). Nevertheless, the members of all four MA Working Groups, including the Sub-Global Working Group, had to be initiated into this way of thinking as the price of their membership. So these were the terms in which we had to think about the meaning of 'pressure' and 'peril' as qualities of small island ecosystems and communities, if indeed we could distinguish them from other coastal ecosystems and communities.

Climate Change Discourse

In retrospect, the MA might be regarded as a momentary diversion from the ongoing labours of the Intergovernmental Panel on Climate Change (IPCC), which is organised in much the same way and subscribes to a similar version of the pressure–state–response model of relationships between human beings and their natural environment. The key difference is that the MA had a particular focus on the relationship between ecosystem services and human well-being, and made greater allowance for the possibility that this relationship could be affected by all sorts of pressures or 'drivers', aside from those directly related to climate change. The IPCC has not been ignoring these other pressures, but is very much concerned to measure the extent to which they add to the measurable impact of climate change on different types of ecosystems or social–ecological systems.

Discussion of this question has been the province of 'Working Group II' since the first assessment was undertaken in 1990. The current mandate of this group is not only to assess the impacts of climate change on different types of natural and human systems, but also to assess their relative vulnerability and their capacity to adapt to these impacts. The first report of Working Group II included a brief discussion of the vulnerability of 'small island countries', especially 'coral atoll nations', to rising sea levels, in a chapter that dealt with 'world oceans and coastal zones' (Tegart et al. 1990: 6-2). The second report of Working Group II had a chapter that dealt with 'coastal zones and small islands' as a hybrid category, noting that small islands could, by some definitions, be regarded as a subset of coastal ecosystems (Bijslma et al. 1996: 293). This chapter also observed that there had been an increasing focus on the vulnerability of small islands because rising sea levels might make them uninhabitable and result in the displacement of people from 'several small island nations' (ibid.: 299). The existential threat to small islands was said to be compounded by 'their distinct ways of life and possibly even their distinct cultures' (ibid.: 306), but this observation was mainly intended to cast doubt on some of the metrics then being used to assess their vulnerability.

In the third assessment, small islands were no longer treated as a distinctive type of ecosystem, but occupied an intermediate or marginal space between 'coastal zones' and 'marine ecosystems' (McLean et al. 2001). On the other hand, they were treated as a distinctive type of *political* system, because a separate chapter was reserved for 'small island states'. This was a reflection of the role being played by the Alliance of Small Island States in global climate change negotiations, although this body had already been in existence since the first assessment was undertaken. Within these states were small island 'communities' whose members had 'legitimate concerns about their future, based on observational records, experience with current patterns and consequences of climate variability, and climate model projections' (Nurse et al. 2001: 855). Those concerns were said to be grounded in their 'high vulnerability and low adaptive capacity', but it was not entirely clear whether these were qualities that were thought to set them apart from other coastal ecosystems or communities. Part of the problem here was the absence of any clear definition of what constituted a small island state or a small island community. A list of small island states in that chapter was a list of the states that were members of the Alliance. It therefore included Papua New Guinea (PNG), even though PNG contains half of the world's second biggest island, while Indonesia, which contains the other half, is not on the list, despite the probability that it contains more small island *communities* than any other nation state.

The subsequent reports of Working Group II have not had a separate chapter on 'small island states', but have instead contained a chapter on 'small islands' that seems to elide the distinction between states, communities and ecosystems (Mimura et al. 2007; Nurse et al. 2014; Mycoo, Wairiu

7

et al. 2022). The fifth assessment report noted that small islands are 'heterogeneous in geomorphology, culture, ecosystems, populations, and hence also in their vulnerability to climate change' (Nurse et al. 2014: 1635), but no attempt has been made to classify small islands according to their relative size. Instead, islands or island states that are obviously very small are cited as extreme examples of the vulnerability that is characteristic of the larger category to which they belong. And that larger category is sometimes represented as one that contains all the *coastal* communities within a set of small island states, as when the latest assessment report observes that 90 per cent of the inhabitants of Pacific Island states—aside from PNG—live within 5 kilometres of a coastline. Rather than seeking to be more specific about the relationship between island size and island vulnerability to the various manifestations of climate change, successive assessment reports have placed a growing emphasis on the significance of local or indigenous 'culture', rather than the actions of small island governments, in framing the adaptive capacities or responses of small island or coastal communities.

The discovery of 'culture' as a variable fact of life on small islands does not constitute a radical departure from the fundamental concern of all IPCC assessment reports with the measurement of things that can be measured. Otherwise, the science of climate change would not be credible as science. Culture is invoked as one of a range of institutional factors that might help to explain why metrics of vulnerability, as applied to small island ecosystems or communities, are unable to predict the way that islanders (or their governments) respond to the measurable impacts of climate change (Adger et al. 2009; O'Brien 2009; Barnett and Campbell 2010; Nunn et al. 2017; Neef et al. 2018).

The absence of any attempt to measure the relative size of different islands, or the extent of their physical isolation from other islands, or the extent of the relationship between these measures and other measures, is best understood as an absence dictated by political considerations. An assessment of the relative size or relative isolation of small islands within each small island state would simply detract from the assumption that these states share a common set of environmental problems and might therefore discover a common approach to their resolution.

The Cultural Critique of Insularity

The appearance of indigenous culture or knowledge as a distinctive but elusive variable in the scientific assessment of climate change on small islands, whatever their actual size, reflects another kind of political consideration that initially had less to do with climate change than with the very concept of insularity. In the Pacific regional context, the critique of this concept was inspired by Epeli Hau'ofa's essay on 'Our Sea of Islands', first published in 1993. Hau'ofa argued that the 'smallness' attributed to Pacific Island states or communities was simply an act of 'belittlement' that denied their capacity to do anything about the problems with which they were confronted. They were seen to be 'much too small, too poorly endowed with resources, and too isolated from the centres of economic growth for their inhabitants ever to be able to rise above their present condition of dependence on the largesse of wealthy nations' (Hau'ofa 1994: 150). But this perception of insularity was simply the effect of a colonial and post-colonial assumption that the ocean itself was not an integral part of the lives and livelihoods of the islanders who lived in its midst, and that assumption had been false since the islands were first occupied because 'much of the welfare of ordinary people of Oceania depends on an informal movement along ancient routes drawn in bloodlines invisible to the enforcers of the laws of confinement and regulated mobility' (ibid.: 156). Regardless of their size, Pacific Islands were really just nodal points in an ancient Oceanic network of social relations that ought to be rediscovered and celebrated in contemporary terms. Smallness and remoteness are not physical attributes worthy of measurement; they are simply 'states of mind' (ibid.: 152).

Hau'ofa's alternative vision of Oceania as a world of mobility, rather than insularity, has since been elaborated by a range of other scholars, mostly anthropologists or historians (Nunn 2004; D'Arcy 2006; Hviding and Rio 2011; Thomas 2011). Hau'ofa was not the first scholar to investigate the relationship between (island) 'roots' and (inter-island) 'routes' in the Pacific (Dening 1980; Bonnemaison 1985; Bayliss-Smith et al. 1988; Thomas 1991). Indeed, this has been a recurrent theme in accounts of the original Austronesian colonisation of the Pacific Island region over the last few thousand years (Bellwood 1978; Gosden 1993; Allen and Gosden 1996; Kirch 1997; Spriggs 1997; Allen 2000; Green 2000; Nunn 2003; Torrence and Swadling 2008). But this relationship has since acquired the shape of a contradiction that is characteristic of all islands everywhere, rather like an 'islandic' version of the contradiction between localisation and globalisation, and contradictions like this are not easily resolved by academic arguments that celebrate or lament the triumph of one force over the other (Clifford 1997; Jolly 2001; Hviding 2003; Hay 2006; DeLoughrey 2007; Baldacchino 2008).

Hau'ofa's essay is widely credited as one of the foundation stones in the creation of a distinctive field of 'island studies', perhaps best conceived as a branch of cultural geography whose contemporary focus can be read from the pages of the *Island Studies Journal* and *Shima* ('The International Journal of Research into Island Cultures'), whose first volumes were published in 2006 and 2007 respectively. The field undoubtedly has older roots, as represented by the establishment of the International Scientific Council for Island Development in 1989 and the International Small Islands Studies Association in 1992. But a reading of the papers published in these two journals over the past 15 years does reveal the extent to which island studies have come to be more concerned with questions of definition than questions of measurement, and with the problematic nature of island identities rather than island livelihoods. Indeed, the hallmark of this new field of study, to paraphrase John Donne, is that no island is an island unto itself. The so-called 'relational turn' that puts the routes before the roots, continually deconstructing the concepts of island isolation or insularity, therefore threatens to dissolve the entity by which the field of island studies is defined (Stratford et al. 2011; Hayward 2012; Hay 2013; Pugh 2013, 2018).

In some respects, the relational turn in island studies is a turn away from the cultural turn in assessments made by the IPCC. One advocate of the relational turn has lamented the tendency of 'resilience ethics', as espoused by Working Group II, to 'throw the violence of the Anthropocene back onto the island community, telling islanders that they need to draw upon their rich community resources in order to survive' (Pugh 2018: 104). Yet the question of who has the capacity or the power to sustain or transform island livelihoods is not a question that has a simple and general answer. This sort of question needs to confront the diversity of island communities instead of being swept under a post-colonial theoretical carpet that turns 'real lives and real islands into the bland non-being of abstraction' (Hay 2013: 212).

Hau'ofa's essay can all too easily be read as a seminal contribution to the weaving of this carpet of abstraction. But Hau'ofa was an anthropologist, and his essay can also be read as an attempt to pose a new set of empirical questions about 'real lives and real islands' in a post-colonial world. So what can anthropologists contribute to an understanding of vulnerability or resilience as variable characteristics of islands and island communities that do vary in many other ways, including their size, even if their size turns out to be immaterial?

From my own experience with the MA and the MBCP, I think the first answer lies in our capacity to go beyond (if not to undermine) a specific form of environmental determinism that accompanies the construction of small islands as terrestrial ecosystems that 'pop up' like unwelcome dinner guests in the midst of coral reef ecosystems whose services to the 'global community' supposedly exceed their services to any local community. This is not the type of environmental determinism familiar to anthropologists, in which societies or cultures somehow 'reflect' their natural environments, but a form of technocratic rationality in which 'environmental management', as a type of human activity, is placed beyond the reach of culture, history and politics (Filer 2009). In this form of technocratic rationality, human agency or power is essentially a function of scale. Global actors make big decisions while local actors make small ones. More importantly, perhaps, the small 'community-based' decisions made by local actors can only be decisions about the services provided by the local ecosystems which constitute their own subsistence livelihoods. The 'resilience' of a small island community is then measured by its capacity to formulate and implement an environmental management plan that constitutes the lowest level in a legal and bureaucratic hierarchy in which each ascending layer manifests a closer relationship between sound science and good policy. On this score, it is possible to endorse the arguments made by Hauʻofa and by subsequent advocates of the 'relational turn' in island studies. However, this does not mean that all islands are in the same metaphorical boat.

The form of technocratic rationality that treats all islands, or all small islands, or all small island states as members of a single class of objects or subjects is perfectly understandable, and probably unavoidable, in the world of ecosystem assessment, biodiversity conservation, disaster risk reduction and climate change adaptation. But the 'cultural turn' now evident in the assessments made by the IPCC constitutes a recognition that members of this class do not share a single set of social relationships. The empirical question is how the physical, economic and cultural qualities of islands might be related to each other in ways that cannot be predicted by a form of cultural geography that simply denies or dissolves the differences between these qualities or between the islands that possess them.

So does size matter? Hau'ofa's essay has given rise to an argument that 'smallness' is an attribute that should be banished from island studies because it connotes a state of disconnection, isolation and dependency (Nunn 2004). But his essay has equally given rise to a recognition that these qualities are not necessarily correlated with each other. Once it is recognised that islands and islanders should be understood 'on their own terms', there is no reason why these terms should not include a consideration of the variable relationship between the physical attributes, like relative size and relative isolation, that are not only visible through a cartographic lens but also a matter of everyday experience (Baldacchino 2020). And when it comes to attributes like dependency, vulnerability or resilience, which are not visible through a cartographic lens, anthropologists should be asking how these can be understood as attributes of the social relationships that connect specific islands or islanders to the rest of the world (Christensen and Mertz 2010; Connell 2015).

From a broad reading of the island studies literature, it is possible to discern three main forms of asymmetrical reciprocity that characterise these social relationships, and to subtract each form of asymmetry from the assertion of a single opposition or contradiction between the connectedness of 'routes' and the disconnectedness of 'roots'.

First, there are 'traditional' relationships in which the asymmetry derives from the natural and cultural endowments of different communities in which one or more of the communities in each relationship is a small island community. Oceania is replete with traditional trading networks involving a social division of labour between small islands, however these might be defined, and bigger islands or other coastal communities. The key question is the extent to which these features of their cultural heritage continue to function as part of their 'real lives' after all the material transformations wrought by the colonial encounter.

Second, there are 'modern' relationships between islanders who stay 'at home' on their island and those members of the same island community who have taken up residence elsewhere. This type of relationship has also been subject to a good deal of ethnographic investigation, but rarely with an eye to testing the hypothetical difference between small islands, or smaller islands, and other sorts of places from which the migrants originate. The so-called 'migration, remittance and bureaucracy' model (Bertram and Watters 1985), which was one of Hau'ofa's main targets of criticism, is simply one variant of this particular form of asymmetry, but one that has mainly been articulated as a feature of small island states rather than small island communities.

Finally, there are 'political' relationships between small island communities and a range of alien intruders who seek to transform their islands in one way or another. If anthropologists have some claim to expert knowledge of this form of asymmetry, I would not argue that this displaces or detracts from the expertise of ecologists, economists or geographers, let alone from the 'indigenous knowledge' ascribed to local communities or the terms on which islanders interpret their own conditions of existence. Hau'ofa was inclined to regard all these alien intruders as people of European descent, or as agents of Western imperialism, but this form of colonial or neo-colonial encroachment does not exhaust the whole variety of ways in which small island communities can be subject to interference by outsiders, including the agents of the small island states to which many of them now belong.

The remaining chapters in this volume explore the interaction between different forms of asymmetry in the social relationships that encompass small island communities in Melanesia.

The Contents of this Book

Most of the chapters in this book derive from papers that were first presented in a panel on 'Small Islands in Peril', jointly organised by Simon Foale and myself, at the conference of the Association for Social Anthropology in Oceania in 2014. Our primary aim in hosting this discussion was to investigate the idea that small island communities could be regarded as canaries in the coal mine of sustainable development because of scientific and anecdotal evidence of a common link between rapid population growth, degradation of the local resource base and intensification of disputes over the ownership and use of terrestrial and marine resources. Our broad hypothesis was that the economic and social 'safety valves' that had previously served to break some of the feedback loops between these trends appeared to be losing their efficacy. We were also concerned that debate about economy–society–environment relationships on small islands had already been overtaken by a narrow focus on the problem of climate change, even in the absence of evidence that sea level rise had become the main driver, or even a measurable driver, of the changes taking place in island lives and livelihoods.

Some of the papers presented in that conference panel have since been revised and published elsewhere (e.g. Connell 2016; Mondragón 2018). Those that became the foundations of chapters in this book have been revised and updated to take account of subsequent events or more recent publications relating to the island or the topic under discussion.

The original version of Chapter 2 had a title in which I asked whether all the small island communities are 'in the same boat'. Whatever metaphor might be chosen, the answer is obviously negative. The real question is how they should be classified. In the first part of the chapter, I deal with this question from the point of view of an imaginary government official in a national planning agency who is looking at a map of PNG and assigning measurable variables to the islands that are visible on the map. I apologise in advance for the rather tedious nature of this exposition, but my aim here is to explore the outer limits of this geographical and statistical form of knowledge, even in the knowledge that PNG no longer has a state that sees things in this way. In the second part of the chapter, I adopt a narrative lens that would be more familiar to anthropologists, by investigating the way that stories about the lives and livelihoods of the small island communities figured in two national newspapers over a ten-year period. This kind of content analysis also has its limitations, since it can only support a classification of the things that make small islands newsworthy. In order to overcome this second set of limitations, I delve into a range of additional documentary evidence to flesh out the most peculiar story, and the longer history, of one small island community that featured in this sample of newspaper articles. The chapter concludes with a discussion of the way that these different perspectives might inform the more detailed ethnographic and historical accounts of the challenges faced by particular communities, as exemplified in the remaining chapters of this book.

The original version of Chapter 3 had a particular focus on the way that members of a group of small island communities in PNG's Milne Bay Province dealt with the challenge posed by the national government's decision to impose a nationwide moratorium on the bêche-de-mer fishery in 2009. This chapter has been revised and expanded to place that particular response in the wider context of efforts to persuade islanders to engage in a more sustainable management of this particular fishery and to appreciate the long-term benefits of conserving marine biodiversity values in the territories under their control. Simon Foale and colleagues recount the way that this particular group of small islands became a focus of attention for the Milne Bay Community-Based Coastal and Marine Conservation Project during the brief period in which that project was being implemented, and place this brief moment of attention in the longer trajectory of change in the lives and livelihoods of the islanders. The authors consider a number of ways in which the science of sustainability, or the scientific assessment of 'ecosystem services', can fail to either influence or comprehend the social and economic choices that islanders make in the construction of their livelihoods.

In Chapter 4, Edvard Hviding recounts the history of the small island in Marovo Lagoon in Solomon Islands, where he has conducted fieldwork for many years. This island is only small in what he calls a 'topographical' sense, for in many other ways it has played a large or dominant role in a regional network of social relations for more than 200 years, from the pre-colonial era in the early nineteenth century to the post-colonial present day. Indeed, so far from being an island 'in peril' or 'under pressure', the inhabitants of this island have produced a variety of perils or pressures for members of neighbouring communities. As Hviding puts it in his conclusion,

> [p]ost-colonial Tusu Marovo is at least as large as its pre-colonial version, and the vulnerability or resilience of its village life and diverse economies are, as in the old days, predicated on and generated from circumstances and relationships far beyond the island itself.

I doubt there is any other small island community in either Solomon Islands or PNG that has managed to sustain this kind of hegemonic position over the same period of time.

In Chapter 5, Nancy Lutkehaus portrays a rather different form of resilience. Manam Island, off the north coast of the PNG mainland, has a history and prehistory that has been disrupted, from time to time, by eruptions of the volcano that constitutes the bulk of the island itself. The islanders had a network of social relations, on and off the island, that enabled them to live with this natural hazard for many centuries, but a major eruption that began in 2004 forced their evacuation to 'care centres' on the mainland. Since then, they have been living in a kind of 'limbo', attempting to reassemble their social networks in a new location while wondering if, how and when they might be able to put them back on the island where they originated. At the same time, the government has been making plans to resettle the whole island community on another part of the mainland, but has proven incapable of implementing such a plan, even while it tries to dissuade the islanders from returning to their island. In January 2023, 18 years after the island was evacuated, the governor of Madang Province announced that the resettlement plan was still in the process of being implemented, but the process has been complicated by the fact that some of the islanders want to go back to the island, some want to stay in the coastal care centres and some want to be relocated further inland (Anon 2023). A kind of moral or political hazard has therefore replaced the original natural hazard as a threat to the survival of this island community, and that is what its leaders now have to contend with.

Chapter 6 returns to the place where I first began to think about the questions to which this volume is addressed. This is the Lihir group of islands in New Ireland Province, where the development of a large-scale gold mine on the biggest of the islands has created a peculiar set of pressures and perils for the inhabitants of three very small islands in the same group. While Richard Jackson and I made some attempt to assess these hazards before the mine was developed, Nicholas Bainton has been investigating the actual process of change for the past two decades, so we have combined the knowledge gained from almost 40 years of research, both by ourselves and by other social scientists, to assemble a detailed portrait of the history of the three very small islands that are known collectively in the local language as rocky or stony places. While this can be read as a story about the negative impact of large-scale industrial development on small island communities, it is also a story about the choices that members of these communities make when faced with such pressures.

A brief concluding chapter provides a summary of the significance of the case study material presented in the main body of the volume.

References

Adger, W.N., S. Dessai, M. Goulden and others, 2009. 'Are There Social Limits to Adaptation to Climate Change?' *Climate Change* 93: 335–354. doi.org/10.1007/s10584-008-9520-z

Allen, J., 2000. 'From Beach to Beach: The Development of Maritime Economies in Prehistoric Melanesia.' In S. O'Connor and P. Veth (eds), *East of Wallace's Line: Studies of Past and Present Maritime Cultures of the Indo-Pacific Region*. Rotterdam: A.A. Balkema (Modern Quaternary Research in Southeast Asia 16).

Allen, J. and C. Gosden, 1996. 'Spheres of Interaction and Integration: Modelling the Culture History of the Bismarck Archipelago.' In J. Davidson, G. Irwin, F. Leach and others (eds), *Oceanic Culture History: Essays in Honour of Roger Green*: Special issue of *New Zealand Journal of Archaeology*.

Anon, 2023. 'Pariwa Hopes to Resettle Islanders Soon.' *The National*, 11 January 2023.

Baines, G., J. Duguman and P. Johnston, 2006. 'Milne Bay Community-Based Coastal and Marine Conservation Project: Terminal Evaluation of Phase 1.' Port Moresby: United Nations Development Programme.

Balboa, C.M., 2009. 'When Non-Governmental Organizations Govern: Accountability in Private Conservation Networks'. New Haven: Yale University (PhD thesis).

——, 2013. 'How Successful Transnational Non-Governmental Organizations Set Themselves up for Failure on the Ground.' *World Development* 54: 273–287. doi.org/10.1016/j.worlddev.2013.09.001

Baldacchino, G., 2008. 'Studying Islands: On Whose Terms? Some Epistemological and Methodological Challenges to the Pursuit of Island Studies.' *Island Studies Journal* 3: 37–56.

——, 2020. 'How Far Can One Go? How Distance Matters in Island Development.' *Island Studies Journal* 15: 25–42. doi.org/10.24043/isj.70

Barnett, J. and J. Campbell, 2010. *Climate Change and Small Island States: Power, Knowledge and the South Pacific*. London: Earthscan. doi.org/10.4324/978184 9774895

Bayliss-Smith, T., R. Bedford, H. Brookfield and M. Latham (eds), 1988. *Islands, Islanders and the World: The Colonial and Post-Colonial Experience of Eastern Fiji*. Cambridge: Cambridge University Press.

Bellwood, P., 1978. *Man's Conquest of the Pacific*. London: Collins.

Bertram, I.G. and R.F. Watters, 1985. 'The MIRAB Economy in South Pacific Microstates.' *Pacific Viewpoint* 26: 489–519. doi.org/10.1111/apv.263002

Bijlsma, L. and others, 1996. 'Coastal Zones and Small Islands.' In R.T. Watson, M.C. Zinyowera and R.H. Moss (eds), *Impacts, Adaptations and Mitigation of Climate Change: Scientific-Technical Analysis*. Cambridge: Cambridge University Press.

Bonnemaison, J., 1985. 'The Tree and the Canoe: Roots and Mobility in Vanuatu Societies.' *Pacific Viewpoint* 26: 39–62. doi.org/10.1111/apv.261003

Bourke, R.M. and T. Betitis, 2003. 'Sustainability of Agriculture in Bougainville Province, Papua New Guinea.' Canberra: The Australian National University, Research School of Pacific and Asian Studies, Department of Human Geography.

Christensen, A. and O. Mertz, 2010. 'Researching Pacific Island Livelihoods: Mobility, Natural Resource Management and Nissology.' *Asia Pacific Viewpoint* 51: 278–287. doi.org/10.1111/j.1467-8373.2010.01431.x

Clifford, J., 1997. *Routes: Travel and Translation in the Late Twentieth Century.* Cambridge: Harvard University Press.

Connell, J., 2015. 'Vulnerable Islands: Climate Change, Tectonic Change, and Changing Livelihoods in the Western Pacific.' *The Contemporary Pacific* 27: 1–36. doi.org/10.1353/cp.2015.0014

——, 2016. 'Last Days in the Carteret Islands? Climate Change, Livelihoods and Migration on Coral Atolls.' *Asia Pacific Viewpoint* 57: 3–15. doi.org/10.1111/apv.12118

D'Arcy, P., 2006. *The People of the Sea: Environment, Identity and History in Oceania.* Honolulu: University of Hawai'i Press.

DeLoughrey, E.M., 2007. *Routes and Routes: Navigating Caribbean and Pacific Island Literatures.* Honolulu: University of Hawai'i Press.

Dening, G., 1980. *Islands and Beaches: Discourse on a Silent Land, Marquesas 1774– 1880.* Melbourne: Melbourne University Press.

Field, M., 2000. 'PNG's "Singing Islanders" Unite to Save Their Home.' *The National*, 20 November 2000.

Filer, C., 2004. 'The Knowledge of Indigenous Desire: Disintegrating Conservation and Development in Papua New Guinea.' In A. Bicker, P. Sillitoe and J. Pottier (eds), *Development and Local Knowledge: New Approaches to Issues in Natural Resources Management, Conservation and Agriculture.* London: Routledge.

——, 2009. 'The Knowledge Problem in the Millennium Assessment.' In J.G. Carrier and P. West (eds), *Virtualism, Governance and Practice: Vision and Execution in Environmental Conservation.* New York: Berghahn Books.

Filer, C., S. Foale, J. Kennedy and others, 2004. 'Sub-Global Assessment of Coastal, Small Island and Coral Reef Ecosystems in Papua New Guinea: Summary National Report.' Unpublished report to the Millennium Ecosystem Assessment.

Filer, C.S. and R.T. Jackson, 1989. *The Social and Economic Impact of a Gold Mine on Lihir: Revised and Expanded.* Port Moresby: Department of Minerals and Energy, Lihir Liaison Committee (2 volumes).

Gosden, C., 1993. 'Understanding the Settlement of the Pacific Islands in the Pleistocene.' In M.A. Smith, M. Spriggs and B. Fankhauser (eds), *Sahul in Review: Pleistocene Archaeology in Australia and Island Melanesia.* Canberra: The Australian National University, Research School of Pacific and Asian Studies, Department of Prehistory (Occasional Paper 24).

Green, R.C., 2000. 'Lapita and the Cultural Model for Intrusion, Integration and Innovation.' In A. Anderson and T. Murray (eds), *Australian Archaeologist: Collected Papers in Honour of Jim Allen*. Canberra: Coombs Academic Publishing.

Hau'ofa, E., 1994. 'Our Sea of Islands.' *The Contemporary Pacific* 6: 148–161.

Hay, P., 2006. 'A Phenomenology of Islands.' *Island Studies Journal* 1: 19–42. doi.org/10.24043/isj.186

——, 2013. 'What the Sea Portends: A Reconsideration of Contested Island Tropes.' *Island Studies Journal* 8: 209–232. doi.org/10.24043/isj.283

Hayward, P., 2012. 'Aquapelagos and Aquapelagic Assemblages: Towards an Integrated Study of Island Societies and Marine Environments.' *Shima* 6: 1–11.

Hoegh-Guldberg, O., H. Hoegh-Guldberg, D.K. Stout and others, 2000. 'Pacific in Peril: Biological, Economic and Social Impacts of Climate Change on Pacific Coral Reefs.' Suva: Greenpeace.

Hviding, E., 2003. 'Contested Rainforests, NGOs, and Projects of Desire in Solomon Islands.' *International Social Science Journal* 55: 539–554. doi.org/10.1111/j.0020-8701.2003.05504003.x

Hviding, E. and K. Rio (eds), 2011. *Made in Oceania: Social Movements, Cultural Heritage and the State in the Pacific*. Wantage: Sean Kingston Publishing.

Jolly, M., 2001. 'On the Edge? Deserts, Oceans, Islands.' *The Contemporary Pacific* 13: 417–466. doi.org/10.1353/cp.2001.0055

Kinch, J., 2001. 'Social Evaluation Study for the Milne Bay Community-Based Coastal and Marine Conservation Program.' Port Moresby: Conservation International (unpublished report to the United Nations Development Programme).

Kirch, P.V., 1997. 'Microcosmic Histories: Island Perspectives on "Global" Change.' *American Anthropologist* 99: 30–42. doi.org/10.1525/aa.1997.99.1.30

MA (Millennium Ecosystem Assessment), 2003. *Ecosystems and Human Well-Being: A Framework for Assessment*. Washington DC: Island Press.

Macintyre, M. and S. Foale, 2004. 'Global Imperatives and Local Desires: Competing Economic and Environmental Interests in Melanesian Communities.' In V.S. Lockwood (ed.), *Globalization and Culture Change in the Pacific Islands*. Upper Saddle River: Pearson Prentice Hall.

McLean, R.F., A. Tsyban and others, 2001. 'Coastal Zones and Marine Ecosystems.' In J.J. McCarthy, O.F. Canziani and others (eds), *Climate Change 2001: Impacts, Adaptation and Vulnerability*. Cambridge: Cambridge University Press.

Mimura, N., L. Nurse and others, 2007. 'Small Islands.' In M. Parry, O. Canziani, J.P. Palutikof and others (eds), *Climate Change 2007: Impacts, Adaptation and Vulnerability*. Cambridge: Cambridge University Press.

Mondragón, C., 2018. 'Forest, Reef and Sea-Level Rise in North Vanuatu: Seasonal Environmental Practices and Climate Fluctuations in Island Melanesia.' In D. Nakashima, I. Krupnik and J.T. Rubis (eds), *Indigenous Knowledge for Climate Change Assessment and Adaptation*. Cambridge: Cambridge University Press.

Mycoo, M., M. Wairiu and others, 2022. 'Small Islands.' In H.-O. Portner, D.C. Roberts, M. Tignor and others (eds), *Climate Change 2022: Impacts, Adaptation and Vulnerability*. Cambridge: Cambridge University Press.

Neef, A., L. Benge, B. Boruff and others, 2018. 'Climate Adaptation Strategies in Fiji: The Role of Social Norms and Cultural Values.' *World Development* 107: 125–137. doi.org/10.1016/j.worlddev.2018.02.029

Nunn, P.D., 2003. 'Nature–Society Interactions in the Pacific Islands.' *Geografiska Annaler—Series B, Human Geography* 85: 219–229. doi.org/10.1111/j.0435-3684.2003.00144.x

———, 2004. 'Through a Mist on the Ocean: Human Understanding of Island Environments.' *Tijdschrift voor Economische en Sociale Geografie* 95: 311–325. doi.org/10.1111/j.1467-9663.2004.00310.x

Nunn, P.D., J. Runman, M. Falanruw and R. Kumar, 2017. 'Culturally Grounded Responses to Coastal Change on Islands in the Federated States of Micronesia, Northwest Pacific Ocean.' *Regional Environmental Change* 17: 959–971. doi.org/10.1007/s10113-016-0950-2

Nurse, L.A., R.F. McLean and others, 2014. 'Small Islands.' In V.R. Barros, C.B. Field, D.J. Dokken and others (eds), *Climate Change 2014: Impacts, Adaptation, and Vulnerability*. Cambridge: Cambridge University Press.

Nurse, L.A., G. Sem and others, 2001. 'Small Island States.' In J.J. McCarthy, O.F. Canziani and others (eds), *Climate Change 2001: Impacts, Adaptation and Vulnerability*. Cambridge: Cambridge University Press.

O'Brien, K.L., 2009. 'Do Values Subjectively Define the Limits to Climate Change Adaptation?' In W.N. Adger, I. Lorenzoni and K. O'Brien (eds), *Adapting to Climate Change: Thresholds, Values, Governance*. Cambridge: Cambridge University Press.

Pugh, J., 2013. 'Island Movements: Thinking with the Archipelago.' *Island Studies Journal* 8: 9–24. doi.org/10.24043/isj.273

——, 2018. 'Relationality and Island Studies in the Anthropocene.' *Island Studies Journal* 13: 93–110. doi.org/10.24043/isj.48

Spriggs, M., 1997. *The Island Melanesians*. Oxford: Blackwell Publishers.

Stratford, E., G. Baldacchino, E. McMahon and others, 2011. 'Envisioning the Archipelago.' *Island Studies Journal* 6: 113–130. doi.org/10.24043/isj.253

Tegart, W.J.M., G.W. Sheldon and D.C. Griffiths (eds), 1990. *Climate Change: The IPCC Impacts Assessment*. Canberra: Australian Government Publishing Service.

Thomas, N., 1991. *Entangled Objects: Exchange, Material Culture, and Colonialism in the Pacific*. Cambridge: Harvard University Press.

——, 2011. *Islanders: The Pacific in the Age of Empire*. New Haven: Yale University Press.

Torrence, R. and P. Swadling, 2008. 'Social Networks and the Spread of Lapita.' *Antiquity* 82: 600–616. doi.org/10.1017/S0003598X00097258

Van Helden, F., 2001. '"Good Business" and the Collection of "Wild Lives": Community, Conservation and Conflict in the Highlands of Papua New Guinea.' *Asia-Pacific Journal of Anthropology* 2(2): 21–44. doi.org/10.1080/1444221011 0001706095

——, 2004. '"Making Do": Integrating Ecological and Societal Considerations for Marine Conservation in a Situation of Indigenous Resource Tenure.' In L.E. Visser (ed.), *Challenging Coasts: Transdisciplinary Excursions into Integrated Coastal Zone Development*. Amsterdam: Amsterdam University Press. doi.org/10.1017/9789048505319.006

——, 2009. '"The Report was Written for Money to Come": Constructing and Reconstructing the Case for Conservation in Papua New Guinea.' In J.G. Carrier and P. West (eds), *Virtualism, Governance and Practice: Vision and Execution in Environmental Conservation*. New York: Berghahn Books.

2

On the Classification of Small Island Communities in Papua New Guinea

Colin Filer

Introduction

Let us assume, for the sake of argument, that an island is a piece of land, or a place on Earth, that is surrounded by water. Let us also assume that an island becomes 'small' when the place in question has a surface area of less than 100 square kilometres. This means that one can probably walk all the way around the edge of it in the space of a single day—or maybe a couple of days if the going is rough or one does not care to walk in the dark. Papua New Guinea (PNG) has hundreds of islands that fit this definition, and dozens of small island communities whose members either live on these islands or think that they belong to them. My aim in this chapter is not to explore the diversity of small islands as physical objects, but to reflect on the diversity of small island communities as human subjects. My more particular aim is to explore the variety of 'pressures' or 'perils' that shape and constrain the livelihoods of small island communities in PNG.

To simplify this task, I take no account of the many small islands located in freshwater lakes or wetlands, since most of these islands contain a very small part of the land used by the people to whom they belong. But if a small island is now defined as a small piece of land surrounded by saltwater, there is equally no reason to assume that small island communities share a single set of features, or a single set of problems, that distinguish them from coastal communities on bigger islands. It can only be argued that small island communities, as here defined, are bound to be coastal communities because small island livelihoods entail the consumption of a mixture of marine and terrestrial resources. So small island communities are a relatively well-defined subset of all the coastal communities whose resident members share this pattern of consumption. If there is one good reason to say that all the people who live on small islands are in the same metaphorical boat, it is that access to real, physical boats—especially, nowadays, fibreglass banana boats with outboard motors—is an important feature of small island livelihoods. So this is an issue, and potentially a problem, for nearly everyone who lives on a small island, or wants to reach one, and may be less of an issue for people who reside in other coastal settlements.

Anthropologists have written detailed accounts of the institutions, practices and beliefs of the people who belong to one or other of PNG's small island communities. These accounts vary in the extent to which they focus on the pressures and perils associated with life in an insular space. The problem of insularity features most prominently in the ethnographic studies of what I shall here call 'miniscule' islands, with a surface area of less than one square kilometre, and it does so because anthropologists have been attracted to miniscule islands with very high population densities (Epstein 1969; Macintyre 1983; Carrier and Carrier 1991; Pomponio 1992; Moyle 2007; Feinberg 2009; Schneider 2012; Rasmussen 2015; Kinch 2020). So these accounts convey a clear sense that population pressure on a limited bundle of natural resources is one of the perilous features of life on a small island. It is not hard to see why this impression should be less evident or entirely absent from ethnographic studies of small islands that are not so small and not so densely populated (e.g. Fortune 1963; Hogbin 1970; Munn 1992; Smith 1994; Foster 1995; Lutkehaus 1995). While it might be possible to extract an argument about the relationship between small island communities and small island ecosystems from a comparison of the studies that focus on miniscule islands with high population densities, this would be a problematic exercise for two reasons: first because the sample itself is so small, and second because the anthropologists have applied quite different conceptual frameworks or theoretical perspectives to their understanding of the relationship. It is therefore difficult to extract a nationwide cultural or political ecology of small island communities from a review of the ethnographic literature.

In this chapter, I shall address the problem of classification in a rather different way. First, I shall address the problem from a fairly conventional geographical point of view, which involves an assessment of the rather limited amount of statistical information that is available from existing documentary records. In the second part of the paper, I investigate a collection of newspaper stories about small islands to see what they can tell us about the place these islands occupy in PNG's national imagination. In the third part, I use a wider range of documentary evidence to tell one strange story about one small island, not because this small island community is in any way typical of the whole set, but rather because it reveals the depth of the problem. I conclude with a discussion of the obstacles that stand in the way of a solution to the problem of classification from any particular disciplinary perspective.

Cartography, Demography, Uncertainty

When viewed through a cartographic lens, PNG consists of a mainland, which is actually half of the island of New Guinea, the second biggest 'island' in the world, and a collection of hundreds of other islands that are contained within the territorial boundaries of the nation state. It is not possible to determine the exact number of these other islands, because some of them are too small to be visible at a certain cartographic scale, while others are so close to each other, or so close to the mainland, that their status as discrete entities is questionable. It is also possible for new islands to appear, or for existing islands to disappear, with the passage of time. There is even greater uncertainty about the number of islands that are occupied or used by people at any particular moment in time. And the smaller the island, the greater the uncertainty.

The Pattern of Islands

What we can do is divide the islands that appear to be occupied at the time of one national census between different parts of the country in accordance with their relative size. Figure 2.1 shows the current division of PNG between 20 provinces and the Autonomous Region of Bougainville (formerly North Solomons Province).

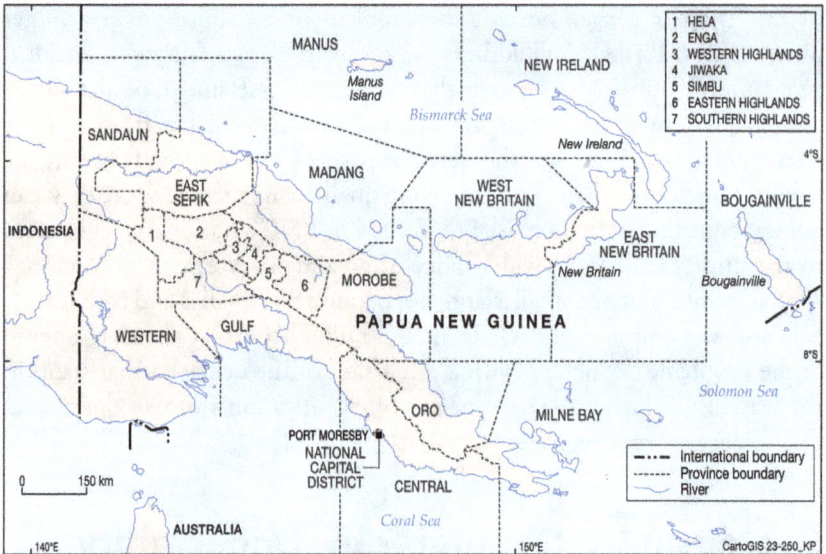

Figure 2.1: Provinces of Papua New Guinea
Source: CartoGIS Services, College of Asia and the Pacific, The Australian National University.

Tables 2.1 and 2.2 show the distribution of populated islands of different sizes between seven regions containing one or more of the provinces that have a coastline. Western, Gulf and Central provinces, along with the National Capital District, are grouped together in what I shall call the Papuan south coast region, while Morobe, Madang, East Sepik and Sandaun (or West Sepik) provinces are grouped together in what I shall call the New Guinea north coast region. An island is here counted as being 'populated' if people were recorded as its residents, either at the time of the 1980 census or at the time of the 2000 census, which was the last census to produce reasonably accurate population figures. The population figures shown in Table 2.2 are taken from the 2000 census.

In this tabulation, island sizes are distinguished on a logarithmic scale. A 'very big' island is more than 10,000 km² but less than 100,000 km² in extent; a 'just big' island is more than 1,000 km² but less than 10,000 km²; a 'medium-sized' island is more than 100 km² but less than 1,000 km²; and a 'small' island is less than 100 km². Small islands can themselves be subdivided along the same lines but, in this instance, the categories are not mutually exclusive since there is no terminology by which they can be clearly distinguished from each other. Small islands are therefore understood to include all the 'very small' islands that are less than 10 km² in extent, which in turn include all the 'miniscule' islands that have a surface area of less than 1 km².

Table 2.1: Distribution of populated islands between regions of PNG

Region	Very big	Just big	Medium	Small
Papuan south coast	0	0	3	28
Milne Bay Province	0	2	8	69
New Guinea north coast	0	0	3	34
Manus Province	0	1	1	42
New Ireland Province	0	2	5	59
New Britain provinces	1	0	1	50
Bougainville Region	0	1	1	25
TOTAL	1	6	22	307

Source: PNG 2000 national census.

Table 2.2: Distribution of small islands and islanders between coastal regions of PNG in 2000

Region	Small		Very small		Miniscule	
	N	Pop	N	Pop	N	Pop
Papuan south coast	28	31,029	16	25,024	10	8,991
Milne Bay Province	69	26,498	58	14,898	20	2,162
New Guinea north coast	34	29,979	25	14,180	19	11,254
Manus Province	42	15,080	38	10,638	30	6,534
New Ireland Province	59	18,748	48	9,063	26	2,919
New Britain provinces	50	38,600	45	16,664	27	8,807
Bougainville Region	25	13,323	23	8,483	17	4,497
TOTAL	307	173,257	253	98,950	149	45,164

Source: PNG 2000 national census.

The one very big island shown in Table 2.1 is the island of New Britain, which has a surface area of approximately 36,000 km² and is divided between the provinces of East and West New Britain. The other six big islands are Fergusson and Normanby in Milne Bay Province, New Ireland and New Hanover in New Ireland Province, and the main islands of Manus Province and the Autonomous Region of Bougainville.

There are only two collections of islands in PNG that cartographers have routinely recognised and named as 'archipelagos'. The Bismarck Archipelago was named after the German Chancellor when the colonial territory of German New Guinea was created in 1884. It is generally understood to comprise all of the islands in what is now the Islands Region of PNG, except for those that belong to what is now the Autonomous Region of Bougainville. This means that it includes all of the islands assigned to the Manus, New Ireland and New Britain regions in Tables 2.1 and 2.2. It has also been taken to include some of the islands assigned to the New Guinea north coast region, like the Siassi Islands in Morobe Province, Long Island in Madang Province or the Schouten Islands in East Sepik Province.

The Louisiade Archipelago in Milne Bay Province is much smaller in scale, but still accounts for most of the islands in Milne Bay Province because it includes most of those that are very small, many of which appear in census records to be uninhabited. It includes four of the province's medium-sized islands (Basilaki, Misima, Sudest and Rossel), which look like the main stepping stones in a chain of islands that stretches eastwards from the eastern tip of the New Guinea mainland. The French navigator who named the island of Bougainville after himself in the eighteenth century named this chain of islands after his king, Louis XV.

Neither of these cartographic entities has any political significance except as a testimony to the legacy of European exploration and colonisation. Louisiade is the name of a local government area in Milne Bay Province, but this only includes half of the islands in the archipelago of the same name. Smaller clusters of islands, inside or outside one of these two archipelagos, were also given European names that have variable degrees of political currency in the contemporary nation state. The Trobriand Islands in Milne Bay Province are generally known as such, but Papua New Guineans hardly ever refers to the main cluster of islands in Manus Province as the Admiralty Islands. They just use the name of the province to designate most of the islands within it.

All of the smaller islands, and even some of the bigger ones, have one or more vernacular names bestowed upon them by their own inhabitants or the inhabitants of neighbouring islands. Some of these individual islands also have European names, mostly dating from the nineteenth century, but these names appear to be losing currency when a vernacular option is available. For example, the island called New Hanover by the German colonial authorities is now more commonly known as Lavongai. Where European and vernacular names for the same island or groups of islands both still have some currency, it makes sense to bracket the second preference after the first. The second biggest island in New Ireland Province is thus Lavongai (or New Hanover), while the most famous group of small islands in Bougainville is better known as the Carteret (or Tulun) Islands.

As previously noted, the names that cartographers use to designate groups of islands may also be present (or absent) in the names of the political or administrative entities that constitute the state itself. The Australian colonial administration was inclined to group small islands together in distinctive 'census divisions' or local council areas. This practice was retained when census divisions disappeared and local government boundaries were changed in 1997.

The Carteret Islands were thus grouped together with four other atoll formations to form what was rather unimaginatively known as the Islands Census Division, and which was then divided between the Atolls Local-Level Government (LLG) area, containing four of these atoll formations, and the Nissan LLG area, which is named after the biggest island in the fifth formation, the Green Islands. These five groups of islands account for more than half of the populated small islands in Bougainville, and most of those that appear to be uninhabited, but they have little in common apart from being atolls and being separated from all the other islands in the Bougainville region by great expanses of water. In practice, each of the five groups is a political entity in its own right. There is another group of small islands in Bougainville, which are not atolls but lie close to the shores of Buka, a medium-sized island whose name is also the name of the LLG area that includes these five small islands.

In the rest of the Islands Region, there are six LLG areas that consist entirely of small islands, and another four that consist of one or two medium-sized islands and a number of small islands in fairly close proximity. In the New Guinea north coast region, there is one LLG area in each of these two categories, and in Milne Bay Province, there are five more LLG areas in the second category. At a higher level of political organisation, there are two districts in which more than one-third of the population counted in the 2000 census were resident on small islands. These are Samarai–Murua District in Milne Bay Province, which encompasses the Louisiade Archipelago, and Manus District, which is the only district in Manus Province. There are three other districts in which small islanders accounted for more than 10 per cent of the population. These are Wewak District in East Sepik Province, Namatanai District in New Ireland Province, and North Bougainville District, which includes the Atolls, Nissan and Buka LLG areas.

Demographic Uncertainties

The organisation of the state, which includes (or should include) the periodic conduct of a national census, is unable to take account of a multitude of uncertainties that surround the distribution of the human population between hundreds of small islands. That is why the numbers shown in Table 2.2 need to be treated with some caution. An exploration of the sources of uncertainty can help to reveal the variety of small island livelihoods while at the same time blurring the boundaries and figures visible to the cartographic eye.

Take the question of who lives where. Government officials have generally tried to limit the number of points at which people are enumerated in order to save time when a census is being conducted. If a small island community consists of households whose members make different uses of the resources available on a number of very small or miniscule islands, this range of activities will be rendered invisible if the census is conducted on just one of these islands, even if all members of the community turn up to be counted. This type of distortion is most likely to take place when islands are clumped together in a major river delta, like the deltas of the Fly and Kikori rivers in the Papuan south coast region, or when they count as components of an atoll formation. Deltaic islands are notorious for shape-shifting in any case, while the number of islands in an atoll formation may vary with the height of the tide (Löffler 1977).

Atoll formations are not all that common in PNG, but they are remarkable for the discrepancy between the number of islands that they contain and the number that were said to be populated in the 2000 census. Aside from the five populated atolls in the Bougainville region, there were three in Milne Bay Province and two more in Manus Province. The combined population of all ten was 11,606, which means that they accounted for less than 7 per cent of the country's small island population. Seventy per cent of these people were living on the Bougainvillean atolls, 20 per cent were in Milne Bay, and only 10 per cent in Manus. The two atolls in Manus are known as the Ninigo and Hermit groups, and they are grouped together in the Nigoherm LLG area. Wikipedia informs us that there are 31 islands in the Ninigo group, while the census tells us that eight of these islands were inhabited by 972 people divided between five council wards. Wikipedia also informs us that there are 17 islands in the Hermit group, while the census tells us that only one of them—Luf—was inhabited, and its population of 158 had a council ward all to themselves. The accuracy of these numbers could be questioned on several grounds, but they do alert us to the likelihood that the total number of very small islands in PNG could be three or four times the number that are supposedly inhabited, while the number that are actually occupied or used at any one moment of time could also be larger than the number where census-takers undertake their enumerations. We cannot be sure whether the resident population of the Hermit group was fully enumerated in 2000, and can only wonder at the frequency with which their local councillor would attend meetings of the Nigoherm council, given the distance between the two island groups.

If these forms of numerical uncertainty are most acute in sparsely populated atoll formations, they are not entirely absent from other parts of the country. For example, the well-known but very small island of Dobu in Milne Bay Province failed to make any appearance in the 2000 census figures, although it probably had about 1,500 residents at the time.[1] Residents of the Tami Islands in Morobe Province, which are also quite well known, are routinely enumerated at a place called Tamigidu, which is located on the coast of the New Guinea mainland, and it is hard to tell how many of the 884 people enumerated at this location were normally resident on the islands. There are a number of other coastal villages on the New Guinea mainland or one of the larger islands where some of the people counted in the census were almost certainly living on small islands nearby, but these smaller islands were not treated as separate census units, so they appear to be uninhabited.

Cases like these can be distinguished from those in which the residents of an island are counted as a distinctive group of people in the census, but it is not clear whether their place of residence should or should not be counted as an island in its own right. This form of uncertainty applies to islands that are now joined to the mainland by means of a causeway, like the 'island' of Tatana in the National Capital District, Los Negros in Manus or Matupit in East New Britain. The national capital even contains a place that was known as 'Koke Island' in the colonial period (Oram 1976: 98) but was never anything more than a small peninsula, and is no longer counted as a separate entity in the national census. Another type of island that may or may not be an island is one that consists of houses built on stilts in a lake or a lagoon. This category would include the two large villages of Wanigela and Waiori, with a combined population of more than 5,000 in 2000. These two settlements are located in Marshall Lagoon and included among the miniscule islands assigned to the Papuan south coast region in Table 2.2.

---

1   This number has been added to the figures shown in Tables 2.1 and 2.2.

Island Alienation

Another type of uncertainty emerges from a comparison between the numbers produced by the 2000 census and those produced by the census conducted in 1980. There are 20 very small islands, 19 of which are miniscule, that were counted as distinctive census units in 2000 but not in 1980. On the other hand, there are 26 small islands, 14 of which are miniscule, that were counted as distinctive census units in 1980 but not in 2000. But this does not mean that 20 islands had been newly settled, while 26 had been deserted, in the intervening period. It could just mean that census units had been reclassified by government officials.[2] Or it could mean, as in the case of Dobu, that the population got counted in one census but were overlooked in the other one.

One pattern that does emerge from this comparison is the apparent depopulation of small islands where the only census units were those designated as 'rural non-villages' in the 1980 census. These account for 18 of the 26 islands that had apparently been deserted by 2000. Most of these islands had been occupied by copra plantations, and the rest by government stations or hotels. The 1980 census counted a total of 31 small islands in which all the residents belonged to a rural non-village, and it is reasonable to assume that these were part of a larger collection of small islands that were alienated from their customary owners or declared to be 'waste and vacant' during the period of colonial administration. European planters, missionaries and government officials were attracted to very small islands, especially in the early colonial period, before the First World War, because they were seen to be places in which there would be less risk of contracting malaria or other tropical diseases. Indigenous Melanesians may well have been attracted to them for the same reason. Some of the islands that were legally alienated were never occupied by Europeans, while others were only occupied for relatively short periods of time, and many had been abandoned by the end of the colonial period. And the process of decolonisation has continued since then. Customary rights have been reasserted over many of these islands—even those that still host townships, suburbs or other settlements that do not count as rural villages (Filer 2014).

---

2   For example, the inhabitants of Kuyawa and Munuwata islands in the Trobriand island group were recognised as the residents of two different islands in the 1980 census, but were recorded as the residents of one of these islands (Kuyawa) in the 2000 census. There is no reason to suppose that Munuwata was no longer inhabited in 2000. On the other hand, the population of M'Buke Island in Manus Province, which was a single census unit in 1980, was divided between two islands and two distinct census units in 2000, and there is no reason to suppose that the second island (Pokali) was uninhabited in 1980.

The miniscule island of Samarai in Milne Bay Province, originally home to the colonial capital of what was once the Eastern Division of the Territory of Papua, was still designated as an urban area in the 1980 census. Twenty years later, its 533 residents were divided between one rural village and one rural non-village, the latter comprising the remnants of the former township. The equally miniscule island of Kwato in the same province, once the local headquarters of the London Missionary Society, was still designated as a rural non-village in 1980. Twenty years later, it contained a single village, with a population of only 73. But this type of transformation did not have uniform demographic outcomes. Garua Island, a very small island in the Bali–Witu group of islands in West New Britain Province, ceased to be a plantation and turned into a single 'village' containing 1,239 people. But Kaleu Island, a miniscule island in the St Matthias group of islands in New Ireland Province, which had also been the site of a plantation, turned into a 'village' that only contained three Papua New Guineans in 2000— presumably members of a single household—who were grouped together with the residents of four other islands in the same group in a single council ward.

Kaleu was the smallest 'village' recorded in the 2000 census, but the fact that it got recorded at all must lead one to wonder how many of the islands that appear to have been deserted or abandoned between 1980 and 2000 were actually still occupied by a few people who either escaped the process of enumeration in the 2000 census or were counted as residents of another village on another island. In those cases where the island had been alienated during the colonial period, one must also wonder whether any remaining residents who would count as 'villagers', if they were counted at all, were the descendants of labourers formerly employed in some colonial enterprise or people who could reasonably claim to be the customary owners of the island in question.

The uncertainty that surrounds the exercise of customary rights to alienated islands is not confined to those islands that have been wholly or partly abandoned by their former occupants. The very small island of Daru was established as the colonial capital of what was once the Western Division of the Territory of Papua at the same time that Samarai was established as the official headquarters of the Eastern Division (Jackson 1976a, 1976b). However, unlike Samarai, Daru was still a provincial capital in 2000, when it had a population of roughly 13,000 people. This meant that it then accounted for more than half of the population resident on very small islands in the Papuan south coast region, and was one of the most densely populated islands in PNG. A recent proposal to develop a new industrial facility on this rather crowded island was quickly greeted with a range of demands from people claiming to be the customary owners of the island (Whiting 2021). It is not clear how many claimants there might be, whether their claims are valid or even whether they reside on the island, but such demands are commonly made, and often meet with some success, whenever proposals are made to develop or redevelop some area of alienated land (Filer 2014).

There is considerable variation in the way that customary rights are recognised or exercised with respect to very small islands in the vicinity of major townships. This point can be illustrated by reference to the four islands in Port Moresby's Fairfax Harbour (Oram 1976; Filer 2019). Tatana Island was never alienated because it was clearly occupied by a substantial indigenous community when the British flag was raised in 1884. Some Tatana people believe that they have customary rights to the miniscule island of Motukea, which was apparently purchased from members of another indigenous community on the adjacent part of the mainland. This island has since been transformed into an international shipping terminal, joined to the mainland by a pair of causeways, and is officially uninhabited because no one actually lives there. Gemo (or Hanudamava) Island is also devoid of human occupants, but has a very different history. This island was apparently purchased from Tatana people, and became the site of a leper colony during the late colonial period, as well as a gun battery during the Second World War. It is now officially regarded as a sort of nature reserve, so Tatana people do not have much to gain from the reassertion of their customary rights. As for Fisherman's (or Daugo) Island, Tatana people have argued, with some success, that it was never legally alienated, despite the assertions of the colonial administration. However, it was occupied by migrants from another coastal Papuan community during the late colonial period, and the Tatana people have not been able to remove them, so it has been alienated in practice regardless of its legal status. There were 752 people living on this island in 2000, which is less than half the number living on Tatana, but the two communities appear to have reached some form of mutual accommodation.

The uncertainties of island alienation are not restricted to this type of peri-urban setting. There is an atoll formation in the Louisiade Archipelago, commonly known as the Conflict group of islands, which had no residents recorded in the 1980 or 2000 census, and had been alienated before the First World War because it appeared to be uninhabited. In 2000, the members of two island communities in the same archipelago were engaged in a dispute over who had customary rights to harvest sea cucumbers from the reefs in this atoll formation, which gave a new meaning to its colonial name (Foale 2005). There were even some plans to give legal recognition to their customary rights if the dispute could be resolved. However, a new tourist resort has since been built on one of the islands in the group, so the process of alienation has been given a new lease of life (Thistleton 2014). In 2022, the Australian businessman who built the resort declared that he was planning to sell the islands to the Chinese Government unless the Australian Government made him a better offer, which prompted cries of alarm from some of PNG's national politicians (Anon 2022a, 2022b; Philemon 2022).

Island Remoteness

The cartographer can see that small islands (or very small islands) vary in their remoteness from bigger islands, when remoteness is simply measured as the distance (in a straight line) from one shoreline to another. The concept of remoteness is problematic when it is taken to entail a kind of isolation that equates to a kind of marginality or even a kind of backwardness. But one place can be a long way from another place without the intervention of a body of saltwater between the two places. People can travel across the sea just as they can travel across the land. The disadvantages of remoteness are not specific to small island communities. They are a function of the need to travel from one place to another and the availability of different means of transportation.

Geographers have tried to calculate the time it takes for people in different parts of PNG to reach a provincial capital or district headquarters (Hanson et al. 2001). These efforts date back to the time when officials in the National Planning Office were constructing a 'provincial data system' back in the 1970s. These measures of accessibility are quite closely correlated with measures of human well-being, like the child mortality rate or the school attendance rate (Filer and Wood 2021). That is not surprising, because these central places contain things like markets, high schools and health centres. But a good deal of uncertainty surrounds the calculations. That is because it is unclear, at any given moment in time, what means of transport are available to the members of a particular community, whether roads are passable or not, how the weather might impact on travelling times, or what public goods and services are actually being delivered at the destination that people are trying to reach.

Table 2.3: Remoteness of small islands from bigger islands or the mainland

Region	< 1 km	1–10 km	10–20 km	20–30 km	30–40 km	40–50 km	> 50 km	TOTAL
Papuan south coast	12	12	2	2	0	0	0	28
Milne Bay Province	8	19	12	9	1	7	13	69
New Guinea north coast	9	12	6	2	1	3	1	34
Manus Province	8	10	2	3	5	1	13	42
New Ireland Province	11	31	5	4	0	0	8	59
New Britain provinces	19	23	2	1	0	0	5	50
Bougainville Region	5	9	0	0	0	0	11	25
TOTAL	72	116	29	21	7	11	51	307

Source: PNG national census maps.

Table 2.3 divides the 307 small islands that were apparently populated in 1980 or 2000 into seven categories that are distinguished by their distance from a bigger island or from the New Guinea mainland. Medium-sized islands have only been counted as 'bigger islands' in this calculation if there is reason to think that they contain a health centre capable of dealing with a medical emergency. Only 12 of the 22 medium-sized islands counted in Table 2.1 are assumed to possess such a facility because of the size of their resident population. The Tabar group of islands in New Ireland Province is unusual in this respect because the local health centre is located on a very small island, Mabua, which is located within 1 km of the shoreline of a medium-sized island, Tatau. Daru Island is likewise unusual because it contains the capital of Western Province, and people from the nearby mainland must therefore travel to this very small island in the event of a medical emergency. Although Daru is more than 1 km from the mainland coast, it is therefore assigned to the least remote category in Table 2.3.

There is no simple way to translate the distances shown in Table 2.3 into quantities of time that it would take for people to reach a health centre, whatever assumptions we might make about the quality of the health services provided at these central locations. If people on a small island need to reach a bigger island in a hurry, the first thing they must normally do is gain access to a fibreglass motor boat and the quantity of fuel required to make the trip. If that can be done, and the weather is not too bad, the speed with which they can travel will be a function of the engine's horsepower and the load of people and cargo in the boat. If the boat travels in a straight line to the nearest point on the shoreline of a bigger island, that point will not normally be the site of the facility that people are trying to reach. So the boat must either travel a good deal further in order to reach that central location, or else the passengers must disembark and travel over land. If the shortest distance between the shores of a small island and a bigger island is somewhere between 30 and 40 km, then a boat might travel that distance in two hours, three hours or four hours, but it might still take a whole day for someone from the small island to reach their destination on the bigger island. And if there is no boat, or no fuel, or the weather is too bad, the trip will not be possible at all.

As the shortest distance between a small island and a bigger island grows longer, so the limitations to travel in small fibreglass motor boats also increase, until a point is reached at which this mode of transport almost ceases to be viable. It is all very well to recall that Pacific Islanders once navigated great distances across the ocean in large sailing canoes, but this mode of transport has been almost entirely abandoned. The people living on the most remote of the small islands in PNG have therefore come to rely on occasional visits by larger boats to supply some of their needs, or enable them to make their way to bigger islands, and are otherwise left to their own devices and resources. All of the populated atoll formations in the Bougainville region are more than 50 km from the medium-sized island of Buka, but some are a lot further away. The distance between Buka and Nukutoa, the one inhabited island in the Mortlock (or Takuu) group, is approximately 240 km. Amotu, the one inhabited island in the Tasman (or Nukumanu) group, is even further away, but is quite a lot closer to some of the bigger islands in Solomon Islands. The very small Western Islands of Manus Province—in the Aua–Wuvulu and Nigoherm LLG areas—are also a very long way from the big island of Manus.

Table 2.4: Population of very small and remote (or far-flung) islands in 2000

	30–40 km		40–50 km		> 50 km		TOTAL	
Region	N	Pop	N	Pop	N	Pop	N	Pop
Papuan south coast	0	0	0	0	0	0	0	0
Milne Bay Province	1	128	7	2,114	12	3,127	20	5,369
New Guinea north coast	1	513	2	1,878	1	68	4	2,459
Manus Province	3	248	1	338	13	2,715	17	3,301
New Ireland Province	0	0	0	0	4	922	4	922
New Britain provinces	0	0	0	0	3	1,315	3	1,315
Bougainville Region	0	0	0	0	10	3,420	10	3,420
TOTAL	5	889	10	4,330	43	11,567	58	16,786

Source: PNG 2000 national census.

There is no simple relationship between the size of populated islands and their relative remoteness. Suppose we make an arbitrary distinction between small islands that are more or less than 30 km from a bigger island that is assumed to contain a functional health centre, or else from the New Guinea mainland. And suppose we say that small islands at a distance of more than 30 km deserve to be called 'remote' or 'far-flung' islands, while those at a distance of more than 50 km qualify as 'very remote' islands. From Table 2.3 it can be seen that 69 out of 307 populated small islands count as remote islands by these criteria, and 51 of them count as very remote islands. Table 2.3 also shows that there are no remote islands in the Papuan south coast region, and that all the remote islands around New Ireland, New Britain and the Bougainville region are very remote. If we then consider the relative size of these small and remote islands, we find that 58 out of the 69 are very small, and roughly half of these 58 very small and remote islands are actually miniscule. In 2000, there were roughly 17,000 people counted as the inhabitants of very small and remote islands, as shown in Table 2.4.

The uncertainty that surrounds the concept of remoteness can be revealed by consideration of the 11 'remote' islands that are small but not very small, and which therefore do not figure in Table 2.4. In the 2000 census, these islands had a combined population of approximately 28,000, but this population was very unevenly distributed between them. There were only 177 people counted as residents of Tong Island in Manus Province, while Unea and Garove, the two main islands in the Bali–Witu group of islands in West New Britain Province, had a combined population of 12,413. Variation in the physical size of these 11 islands is only partially correlated with variation in the size of their population. Some are more densely populated than others. When a 'remote' island has a substantial population, it is more likely to have a functional health centre, and even an airstrip. And if the presence of such things makes the island seem less 'remote' than it did before, it also reduces the 'remoteness' of very small islands in the same island group. From this point of view, the Tanga and Anir island groups in New Ireland Province are no more 'remote' than the Tabar and Lihir island groups. All four island groups are roughly equidistant from the coast of the big island of New Ireland. The difference between them is that the two main islands in the Tabar group (Tatau and Tabar), like the main island in the Lihir group, qualify as medium-sized islands, while the two main islands in the Tanga group (Boang and Malendok), and the two main islands of the Anir group (Babase and Ambitile), do not qualify for this status because their surface area is more than 10 km² but less than 100 km².[3]

We can assess the demographic correlates or consequences of remoteness by calculating the size of the rural communities resident on islands that are either small, very small or miniscule, and then asking whether remoteness makes any difference to the size of the resident population. Table 2.5 shows the results of this calculation. Urban islands like Daru are excluded from this calculation, as are islands whose only occupants were counted as residents of 'rural non-villages' in the 2000 census. The total number of populated small islands under consideration is thus reduced from 307 to 270, while the number of these islands that are very small is reduced from 253 to 216, and the number that are miniscule is reduced from 149 to 127. Where these islands contained a rural village population and a (generally much smaller) 'non-village' population, the 'non-villagers' are also excluded from the count. So Table 2.5 gives a reasonable indication of the size of the indigenous population, or the population of customary landowners, on islands that differ in both size and remoteness.

Table 2.5 tells us that the average size of the rural village population on the 66 islands that are both small and remote was considerably larger than the average size of the rural village population on all of the 270 small islands in this sample. But it also tells us that this relationship was reversed in the case of the 216 islands that are very small and the 127 islands that are miniscule. The very small or miniscule islands that are more remote have a smaller average population than those that are less remote. But these averages mask a wide range of variation.

---

3   The remaining islands in this collection of 'remote' islands that are small but not very small are: Gawa (in Milne Bay Province), Vokeo (in East Sepik Province), Baluan (in Manus Province), and Nissan (in the Autonomous Region of Bougainville).

Table 2.5: Numbers of rural villagers counted as living on small islands in 2000

Island type	< 100	100–499	500–999	1000–1999	> 2000	Total	Mean
All small	36	156	44	23	11	270	570.2
Small and remote	11	37	10	4	4	66	656.8
All very small	35	140	29	10	2	216	368.5
Very small and remote	11	36	7	1	0	55	303.0
All miniscule	28	82	12	3	2	127	323.2
Miniscule and remote	7	20	1	0	0	28	219.1

Source: PNG 2000 national census.

We have already noted the wide range of variation in the size of the resident population on the 11 islands that are remote and small, but not very small, and that variation remains when we subtract the rural non-village population from the total population. The remote islands of Unea and Garove in the Bali–Witu group had village populations of 8,520 and 3,220 respectively. Boang island in the Tanga group had 4,238 rural villagers, while Nissan Island, the main island in the Green Island atoll formation in the Bougainville region had 4,064. Three of the other islands in this category had village populations of more than 1,000 but less than 2,000, three had village populations of more than 500 but less than 1,000, and one (Tong) had just 177 villagers. A similar range of variation is found among the 42 small islands that are neither remote nor very small. At one extreme, Manam Island in Madang Province had a village population of 7,310. At the other extreme, Burusan (or Baudisson) Island, which belongs to the Tigak group of islands in New Ireland Province, has a surface area of more than 10 km², but was home to just 11 rural villagers in 2000.[4]

The two miniscule islands occupied by more than 2,000 rural villagers are so far from being remote that they hardly even count as islands. These are the settlements of Wanigela (with 3,322 residents) and Waiori (with 2,003 residents), which consist of houses built on stilts in the waters of Marshall Lagoon, in immediate proximity to the mainland of Central Province. Aside from these two artificial islands, there were ten very small islands occupied

---

4   Since this island was home to a single rural non-village in the 1980 census, it is most likely an island that was alienated during the colonial period, and the 11 'villagers' counted as the island's only inhabitants in the 2000 census were the families of former plantation labourers.

by more than 1,000 rural villagers. The most populous was probably the island of Dobu in Milne Bay Province, but we do not know how many villagers were living on it because they did not get counted in the 2000 census. Some of the other islands in this category were anomalous in other ways. Of the three that were miniscule, two were essentially peri-urban islands, even though their residents were counted as a rural villagers in the 2000 census. One was Sek, near the township of Madang, the capital of Madang Province. The other was Matupit, joined by a causeway to the township of Rabaul, which was the capital of East New Britain Province until a large part of it was destroyed by a volcanic eruption in 1994. Matupit survived the eruption with all its buildings intact, but the miniscule island of Sissano in Sandaun Province was not so fortunate. Sissano is a sandbar located between the lagoon of the same name and the Bismarck Sea, and barely counts as an island because the eastern end is just about connected to the New Guinea mainland. The very large village of Arop, which was built on this 'island', was completely destroyed by a tsunami in 1998, and many of the villagers were killed, as were many of the residents of nearby coastal villages. Although the 2000 census assigned 1,769 people to this village, it is not clear where they were actually living at the time.

Table 2.5 shows that there was only one very small but remote island that had a village population in excess of 1,000 at the time of the 2000 census. This was Bam (or Biem) Island in the Schouten island group in East Sepik Province. The 1,154 villagers living on this island outnumbered the 896 who were living on the small, but not very small, island of Vokeo (or Wogeo) in the same island group. Table 2.5 also indicates that remoteness alone makes little difference to the range of populations recorded on the 204 very small islands that had less than 1,000 people living on them in the year 2000. The only miniscule and remote island with a population of more than 500 and less than 1,000 was Koil Island, with a population of 724, which also belongs to the Schouten island group. The distribution of the population between these three islands exemplifies a pattern observed in some other island groups, where population density is inversely related to the size of the island. Other miniscule and remote islands with populations in excess of 300 are located in atoll formations, but it is hard to observe a relationship between population density and island size in these formations because many of the islands appear to be completely uninhabited. Even if we discount the five anomalous cases in which miniscule islands appear to have had village populations in excess of 1,000, the figures shown in

Table 2.5 suggest that there were roughly 100 very small islands with population densities in excess of 100 per square kilometre, most of which were miniscule, and some of which were also remote.

Island Communities

The figures shown in Table 2.5 take no account of the relationship between rural village census units, which may themselves be very small or miniscule, and the council wards that constitute the lowest level of representation or participation in PNG's current system of local-level government. In rural areas, the general assumption behind this system is that council ward boundaries should be broadly aligned with the boundaries of traditional (or pre-colonial) political communities. Since the census assigns rural villagers to council wards, as well as to census units, it is therefore possible to ask whether the remoteness of small island populations makes any difference to the size of these neo-traditional political communities.

Of the 36 small islands with a miniscule population of less than 100 rural villagers, only seven were designated as distinctive political communities in this sense. All seven were very small islands, and five of the seven were miniscule. Four of the seven were also remote. Wei Island, yet another island in the Schouten group, and Liot Island, in the Ninigo atoll formation, each had 68 villagers and one local councillor to represent them. Alcester (or Nasikwabu) Island, in Milne Bay Province, had a population of 93, while Tench (or Enus) Island in New Ireland Province had a population of just 66. These last two islands are not only remote from bigger islands, but are a long way from any other populated island of any size at all.

If we then shift our gaze to the larger collection of 156 small islands that had a somewhat larger, but still quite small, population of less than 500 villagers, we find that 49 of them were designated as distinctive political communities in the sense of constituting a single council ward. Sixteen of these 49 island communities were remote, 15 of these remote communities occupied a very small island, and seven of these very small islands were miniscule. There were two other very small islands, both of them remote, whose village population was less than 500, but was divided between two council wards. One was the island of Motorina, in Milne Bay Province, with a population of 455; the other was the island of Aua in Manus Province, with a population of 419. It is notable that all these remote and relatively autonomous island communities with relatively small populations are confined to four of the regions distinguished in Tables 2.1 and 2.2, namely Milne Bay, Manus, New Ireland and Bougainville. And they include a number of the atoll communities in Manus and Bougainville.

Seventy per cent of the 192 small islands with fewer than 500 villagers living on them in 2000 did not count as autonomous political communities, but were designated as census units that were part of a larger council ward. There were another 13 small islands with bigger village populations that were treated in the same way, but only one of them counted as a remote island by the criterion adopted here. The islands whose village populations belong to larger political communities are generally of two kinds. In some cases, the residents of one or more small islands are grouped together with people living in one or more settlements on a bigger island or on the mainland of New Guinea. In most other cases, the people living on a number of small islands in fairly close proximity to each other are grouped together in a single council ward.

A number of the small islands in the New Guinea north coast region exemplify the first arrangement. As previously noted, residents of the Tami Islands in Morobe Province were actually counted among the 884 residents of the council ward known as Tamigidu, which is shown on maps as a pair of mainland census units. Likewise, the 1,005 villagers living on the peri-urban island of Sek in Madang Province were grouped together with 618 villagers living in two mainland census units to form the council ward known as Ward 10.[5] The 624 villagers living on the island of Ali in Sandaun Province were grouped together with 64 residents of a mainland census unit known as Ali Beach, although the council ward to which they all belonged is still known as Ali Island. In this last case, it seems fairly clear that some members of what had traditionally been a single island community had migrated to the mainland and established a sort of beachhead there. Some of the other council wards that have been constituted in this way, including Tamigidu, are likely to have a similar history of population movement, but there could be other cases in which a very small or miniscule island has been occupied by migrants from a larger community that was already established in another location.

---

5   Madang is the only province in PNG where council wards are given numbers rather than proper names.

The most glaring example of the second arrangement is the Nonovaul island cluster in New Ireland Province. The Nonovaul council ward covers 13 small islands in what is commonly known as the Tigak island group because Tigak is the name of the language spoken by the indigenous population. The islands in this group are located between the big island of Lavongai (or New Hanover) and the main island of New Ireland, and their residents are divided between three wards in the Tikana LLG area. Eleven of the islands in the Nonovaul ward are very small, and nine of these are miniscule. In 2000, the total village population of all 13 islands was 1,311. The most populous was the miniscule island of Nonovaul, with 267 residents, from which the island cluster and the council ward derive their names. But seven of the islands in this cluster had fewer than 100 residents.

When the population of one small island is split between a number of local council wards, the number of wards is partly but not entirely correlated with the size of the island's population. We have already noted the existence of two very small islands (Motorina and Aua) that are split between two council wards, despite having fewer than 500 residents in 2000. Of the 44 small rural islands that had somewhat larger village populations, but still fewer than 1,000 residents, eight had two council wards, one had three, and two had four. Of the 34 small rural islands that had even larger village populations, 11 had two wards, two had three, three had four, and six had even more. The record was set by Manam Island in Madang Province, with no less than 14 council wards for a population of 7,310. While it may be true that the boundaries of rural council wards can be viewed as the boundaries of neo-traditional political communities, the average population of these communities, or the range in the size of their population, varies between different parts of the country in ways that have more to do with the current configuration of the state than with pre-colonial forms of political organisation.

This point is most clearly illustrated by the contrast between Manus Province and the Autonomous Region of Bougainville. Under the current constitutional framework, Manus should have only three rural LLG areas, because it only contains one district, but it actually has four times this number. Since the average size of the population represented in each of these LLGs is so much smaller than it is in other parts of the country, the average population of each council ward also tends to be much smaller. So the constitution of the Nigoherm LLG area, which had six wards with a total village population of 1,018 in 2000, is not simply a reflection of the remote location of the Ninigo and Hermit atoll formations, nor the small size of their traditional political communities. It is just an extreme case of the political fragmentation that is characteristic of this particular province.

In 2000, the population of the Nigoherm LLG area was quite a lot smaller than the population of a cluster of six small islands off the northwestern coast of the main island of Bougainville that made up the Islands ward in the Kunua LLG area. Five of these islands are very small, and four of these five are miniscule. The 2000 census counted 1,911 villagers across all six islands, and since most of them share a single vernacular language, the name of this language (Saposa) is sometimes applied to the whole cluster.

It is true that very remote atoll communities, like those that occupied the Mortlock, Tasman or Nuguria islands, were also designated as distinct wards, within the Atolls LLG area, despite having populations of less than 500 at the time of the 2000 census. But they could not possibly function as parts of a single political community, given the physical distance between them and the absence of a common vernacular language. Most of the council wards in the Bougainville region have much larger populations. The very small islands of Petats and Matsungan, close to the western coast of Buka, had a combined population of 1,640 in 2000, but they were still combined with six rural village census units on Buka itself to form a single council ward, called Tonsu, with a total village population of 3,899.

Political and administrative boundaries in the Autonomous Region of Bougainville have been subject to a number of changes over the past 20 years because of the region's partial separation from the nation state of PNG. At the time of the 2000 census, the 'local-level governments' to which council wards were assigned were largely fictions of the state's imagination. If the wards themselves had more substance as political communities, then they resembled the 'local-level governments' in Manus Province that were more like community governments than their counterparts in other parts of PNG. The inference to be drawn from such comparisons is that topographical maps and national census data cannot tell us very much about the number of island communities that existed in 2000, or at any other time, because they cannot tell us how the physical separation of islands is related to the social construction of communities.

Island Migration

Insofar as some small island communities are under pressure because their rate of population growth is accompanied by a growing shortage of land or other natural resources, one might expect that such communities would have higher rates of out-migration than other communities with lower levels of population pressure. There are two ways to test this type of hypothesis with a mixture of cartography and government statistics. The provincial data system (PDS) assembled by staff of the National Planning Office in the 1970s differed from the subsequent national census because it included a count of the number of people who originated from each rural village census unit but were designated as 'absentees' because they were more or less permanently living somewhere else. That was the last time such calculations were made for the whole of PNG. What we can calculate from national census data is the rate of change in the resident population of each census unit from one census to the next—at least in those parts of the country where the names of census units have not changed in the meantime. And even if the names have changed, we can still calculate the rate of population growth on small islands because we have a pretty good idea of which census units belong to which islands. So we can ask whether small islands that had unusually high population densities in 1980 also had unusually high levels of absenteeism in 1979 or unusually low rates of growth in their resident population between 1980 and 2000. We can also ask whether island size or island remoteness made any difference to either of these two dependent variables.

Tables 2.6 and 2.7 summarise the results of such calculations when applied to the second of these questions. Although the total number of small islands in both tables appears to be the same, they are not in fact the same collection of islands, and that is why the other totals in the two tables are not identical. There are a few islands whose village populations (and numbers of absentees) were recorded in the PDS and the 1980 census, but whose village populations escaped enumeration in the 2000 census. Dobu Island is a case in point, but that was clearly a case of administrative oversight. Most of the others were very small islands with very small populations in 1980, and were most likely amalgamated with other island census units in 2000. On the other hand, there is a somewhat larger number of islands, including most of those in the atoll formations of the Bougainville region, for which we do have measures of intercensal population change but which were somehow omitted from the PDS in 1979.

Table 2.6: Percentage of villagers absent from 241 small islands containing rural villages in 1979

Island type	<10%	10–19%	20–29%	30–39%	40–49%	> 50%	Total	Mean
All small	59	82	44	26	21	9	241	20.4
Small and remote	16	25	8	2	2	3	56	16.6
All miniscule	18	29	21	14	16	5	103	24.4
Miniscule and remote	5	8	1	0	2	3	19	21.6

Source: PNG provincial data system, rural community registers.

Table 2.7: Village population growth rates on small islands with rural villages, 1980–2000

Island type	Negative	1–49%	50–99%	100–149%	150–199%	>200%	Total	Mean
All small	19	71	97	37	6	11	241	73.0
Small and remote	5	20	30	1	6	3	65	63.2
All miniscule	10	30	45	15	4	4	108	73.1
Miniscule and remote	4	9	10	4	0	1	28	55.5

Source: PNG national census data.

Table 2.6 suggests that the proportion of absentees in small island communities, as documented in 1979, tended to be greater on miniscule islands than on other small islands, but tended to be lower on remote islands than on islands that are less remote. But how might we explain the very wide range of variation in the absentee rates between islands of the same size or the same degree of remoteness?

Of the 59 small islands where the proportion of absentees was less than 10 per cent in 1979, 23 had an absentee rate of less than 5 per cent, and five had an absentee rate of less than 1 per cent. Milne Bay Province accounted for 27 of the islands with absentee rates below 10 per cent, 13 of those with rates below 5 per cent, and four of the five with rates below 1 per cent. If we confine our attention to miniscule islands, we find that this one province accounted for nine of the 18 with rates below 10 per cent and five of the seven with rates below 5 per cent. This province also contained the only two miniscule islands with absentee rates below 1 per cent. Indeed, these two islands, Konia and Munuwata, with resident village populations of 74 and 128 respectively, had no absentees at all. Konia was the only remote island where absentees accounted for less than 1 per cent of the total village population. So it looks as if Milne Bay is a region where islanders had a peculiar inclination to stay at home.

This impression is confirmed when we consider the geographical distribution of the 30 small islands with absentee rates of 40 per cent or more. Only one of these islands (Bonarua) was located in Milne Bay Province, whereas 12 were located in Manus Province, and the rest were scattered around the other regions shown in Tables 2.1 and 2.2. Given that Manus has a smaller number of small island communities than Milne Bay, New Ireland and New Britain, it would appear that Manus is a region where islanders had a greater inclination to move elsewhere.

While six of the nine small islands with absentee rates above 50 per cent were in Manus Province, the island with the highest absentee rate was Parama, located in the estuary of the Fly River, where 557 people from a total village population of 661 were said to be living somewhere else in 1979. The PDS does not tell us where they were living, but in this particular case, a significant proportion were probably resident on the urban island of Daru, the capital of Western Province, which is not very far away.

But this case also exemplifies another point, which is that absentee rates could sometimes vary quite markedly between islands in one particular part of the country. The island of Dibiri, which is also located in the estuary of the Fly River, though a good deal further from Daru, had only 14 absentees from a total village population of 670, and was, therefore, one of the islands with an absentee rate below 5 per cent. In the Calvados island chain in Milne Bay Province, which is part of the Louisiade Archipelago, the absentee rate varied from a low of 3.5 per cent on Panawina Island (with seven absentees in a population of 198) to a high of 22.4 per cent on Brooker Island (with 71 absentees in a population of 317).[6] In the Ninigo atoll formation in Manus Province, the rate varied from a low of 1.2 per cent on the miniscule island of Liot (with one absentee in a population of 84) to 53.3 per cent on the equally miniscule island of Awin (with 40 absentees in a population of 75).

Table 2.7 suggests that there is no difference between average population growth rates on islands of different sizes, but the growth rate tends to be lower on the more remote islands than on those that are less remote. Nevertheless, we can still observe a very wide range of variations in rates of change between islands of the same size or the same degree of remoteness.

Of the 19 islands whose resident village population appears to have shrunk between 1980 and 2000, one was in Milne Bay Province, four were in Manus Province, and the rest were scattered around the other regions of the country. By far the most populous of these 19 islands was Matupit, whose resident population shrank by more than 20 per cent, from 2,016 in 1980 to 1,599 in 2000. But this was a special case, since we know that many members of this island community were resettled in another part of East New Britain Province after the volcanic eruption of 1994 (Martin 2013). All but one of the other islands with shrinking populations had fewer than 500 residents in 1980. The greatest shrinkage was recorded on the very small island of Butei in the Tigak island group, a census unit in the Nonovaul council ward. This island was occupied by 130 villagers in 1980 but only 51 in 2000.

We might expect this form of depopulation to be associated with high absentee rates in 1979, but there is no obvious correlation between these two variables. Only two of these shrinking island communities had absentee rates above 40 per cent in 1979, while four had absentee rates below

---

6   Brooker is one of the most densely populated islands in the whole of Milne Bay Province.

10 per cent. The absentee rate on Matupit was 20 per cent, on Butei it was 16 per cent, and the average for all of the 19 islands with negative growth rates was 19 per cent, which is lower than the mean absentee rate on all of the 241 small islands documented in the PDS.

The four islands with absentee rates below 10 per cent included the miniscule and remote island of Liot, in the Ninigo atoll formation, whose resident population declined from 83 to 68. Liot was one of eight islands with declining populations that had fewer than 100 residents in 1980, so we might imagine that these islands were in the process of being abandoned by their remaining residents, regardless of the number of former residents who had already left them by 1979. However, eight of the 11 islands where the population appears to have grown by more than 200 per cent also had fewer than 100 residents in 1980. And these remarkable rates of growth were also associated with very different absentee rates in 1979. For example, the miniscule and remote island of Tench in New Ireland Province, which had 21 residents in 1980 and 17 absentees in 1979, boasted 66 residents in 2000. By contrast, the very small and remote island of Luf, the only inhabited island in the Hermit atoll formation, which had 41 residents in 1980 and four absentees in 1979, boasted 158 residents in 2000.

We would not expect a natural rate of population growth to be more than 100 per cent over a 20-year period, so migration must account for the higher rate of increase on the 53 islands whose resident populations more than doubled over this period. No doubt some of the migrants were former absentees. The island of Parama in Western Province, which had the highest absentee rate in 1979, also recorded one of the highest growth rates between 1980 and 2000, as the resident population grew from 104 to 438. This could well be taken to mean that some of the 557 people recorded as absentees in 1979 had returned to the island. But Parama, like Matupit, may be an exceptional case.

Luf was one of 13 islands with seemingly unnatural growth rates that had absentee rates of less than 10 per cent in 1979. Another was the miniscule island of Panamen (or Iyen), part of the Jelewaga council ward in the Louisiade Archipelago, whose resident population increased almost tenfold, from 11 to 116. The average absentee rate for all of the 53 islands with exceptionally high rates of population growth was 22.5 per cent, which is only marginally higher than the mean absentee rate on all of the 241 small islands documented in the PDS. So the homecoming of former absentees cannot be the only factor that accounts for such high rates of increase.

The numbers also suggest a form of migration between very small islands that constitute parts of a larger island community, or else between rural communities in the same part of the country. This at least seems more likely than the adoption of unrelated migrants from places further afield.

However, we also need to make some allowance for the possibility that the numbers cannot be trusted. The PDS might only have recorded the numbers of people who had moved away from an island quite recently, and the way that absentees were counted could have varied between different provinces or districts, depending on the way that government officials went about the business of enumeration. For example, it is possible that officials in Manus Province were more assiduous in recording absentees than their counterparts in Milne Bay. Nor can we be sure that the whole of the resident population of each island or each island community was actually counted in each national census. People who were not counted might have been present, or else they might have been absent for only a short period of time. Half the population of a very small island community could have been away on some kind of expedition at the time a census was conducted. So the rates of population growth derived from a comparison between the 1980 and 2000 census figures at a strictly local level may be statistical illusions— at least in some cases.

Mediated Narratives of Pressure

From what is already known of their geographical and demographic characteristics, we might now be able to divide the small islands of PNG into a number of different 'boats', and then try to make some inferences from existing ethnographic evidence about the pressures and perils to which they are subject. However, this application of the comparative method is liable to assume that such evidence is randomly distributed between islands of different types, which is clearly not the case. My own preference would be to use such evidence in a more strategic way, as a sort of reality check on the stories about small island communities that have been published in the national newspapers or other media outlets over some period of time. Having collected a bundle of such stories since my initial engagement with the Millennium Ecosystem Assessment (see Chapter 1), I was initially surprised by the extent of variation in the degree of notoriety or newsworthiness attached to different islands with similar geographical and demographic characteristics, as well as in the reasons for their notoriety.

This means that we need to consider the factors that contribute to the newsworthiness of a place, or a collection of places, whether or not they be small islands, and how these factors reflect the pressures or perils to which their residents are subject.

A Pattern of Stories

From the digital editions of the two national newspapers that were published over a period of ten years, from the beginning of 2004 to the end of 2013, I downloaded all the items that clearly and specifically covered one or more of PNG's small islands and inserted the whole lot into a single document. I cannot be sure that I managed to identify all the relevant newspaper items, but am fairly confident that I captured more than 90 per cent of them. The 555 items in question included occasional editorials, or letters written to the editors, as well as regular articles, but did not include the so-called 'advertorials' through which people pay to advertise their own opinions, since these are not available in the digital versions of the newspapers. I did not attempt a comparable survey of blog posts or Facebook posts over the same period because these forms of national and local discourse were quite limited in their volume at the beginning of this period, and while their volume has since grown quite rapidly, there is no reason to suppose that their producers have a set of concerns that are distinct from those canvassed in the newspapers. Most of the regular articles published in the newspapers, with or without a byline suggesting authorship by a journalist, are actually based on press releases produced by other individuals with a vested interest in a particular topic, who might be politicians, public servants, or representatives of private companies, aid agencies or civil society organisations. The national newspapers can therefore be counted as a national forum in which these different voices make themselves heard.

My analysis of the content of this decade of discourse about small islands has been conducted by means of a spreadsheet in which each row contains a single 'story' about one or more of the islands. A story may consist of more than one item from the newspapers if it deals with a single topic in a single place over a single period. The size of the story can thus be measured by the number of items (or the number of words) that it contains. On this score, the biggest story of the decade was the displacement of the population of Manam Island by a volcanic eruption on 24 October 1994 and the woeful account of subsequent efforts to resettle them on the north coast of the mainland. This story accounted for 80 items over the course of the decade.

At the other end of the scale, a story could be very small if it contained a single item that made reference to a number of small islands in completely different locations. The smallest story in my sample was one about the visit of a luxury cruise liner to a number of small islands in four different provinces in May 2004 (Anon 2004).

The smaller the story, the easier it is to identify the voice and the interest behind it. The story about the cruise liner was clearly based on a press release issued by the company that owned the boat. It could just as well have been an advertisement. But the story about the consequences of the eruption on Manam Island featured the views of a broad range of 'stakeholders', from the prime minister to the provincial governor, the local member of parliament (MP), provincial government officials, local community leaders, and so forth. And because a big story could take different twists and turns with the passage of time, and different parts of the story could be separated by periods of silence, it can be hard to decide whether it was in fact a single story or a number of different stories with something in common.

Of course, the point of doing this kind of content analysis is not to determine what constitutes a single story, as opposed to a set of stories with something in common, but to assess the relative significance of the things that make them newsworthy in the first place. This can be done by reference to the pressure–state–response model that has come to inform scientific assessments of the relationship between human activity and environmental change. For the Intergovernmental Panel on Climate Change, there is ultimately one big source of pressure, which is the volume of greenhouse gas emissions from a range of human activities, and responses are divided between efforts to reduce or 'mitigate' that pressure and efforts to adapt to its negative effects. For the Millennium Ecosystem Assessment and its successors, there are numerous pressures or drivers of change in the relationship between ecosystem services and human well-being, and these changes give rise to an equally wide range of responses on the part of human decision-makers (MA 2005).

Environmental conditions are not generally newsworthy unless a story can be told about the drivers of environmental change or the measures being adopted to deal with the consequences. This point applies to small island ecosystems as much as it does to other kinds of natural environment. A small part of the story about Manam Island was about the effects of a particular type of natural hazard or disaster, which has no obvious connection to climate change, on the livelihoods of the islanders affected by it. A much bigger part was about a remedial measure—a program of resettlement—that could equally well be adopted as a response to pressures of a different kind, including those associated with climate change. But some of the stories told about small islands have no obvious connection to changes in the relationship between the islanders and their island ecosystems. The visit of a cruise liner is a case in point. The islands sound like a source of attraction to tourists, but the visit might not exactly count as a response to any problem that confronts the islanders, except perhaps a lack of income-earning opportunities.

Most of the items in my collection can be divided between two categories. On one side are those that deal primarily with different types of pressure, peril or threat—what the victims are nowadays inclined to call different forms of 'suffering'. On the other side are those that mainly deal with responses or remedial actions taken by islanders or their allies to solve these problems. There is only a small number of cases in which equal weight is placed on both the problem and the solution, and there are several cases, like the story about the cruise liner, where it is hard to identify any narrative connection between pressure and response, or problem and solution.

Tectonic Disruptions

The Manam Island story began with an item published in February 2004, well before the eruption took place, in which mainland community leaders voiced their opposition to any plan to resettle the islanders on the mainland. That is because an eruption was expected. In the months after it actually happened, there were numerous items about the way that the resettlement exercise was actually implemented, then a period of relative silence, and then a sequence of occasional items, between 2009 and 2013, that were mainly concerned with the conflict between the islanders who had been resettled and their mainland neighbours, which was predictable, and what might be done to resolve that conflict, which was a puzzle without an obvious solution.

This story is part of a broader and longer narrative about the way that seismic events have disrupted the lives and livelihoods of coastal communities, including small island communities, most especially those located in the Bismarck Archipelago. If my collection of stories had started ten years earlier, then the eruption of Mount Tavurvur, which damaged the township of Rabaul and threatened the community of Matupit in 1994, would have been a big story. The tsunami that afflicted the community of Sissano and neighbouring coastal communities in 1998 would have been an even bigger one. The after-effects of those events were still making occasional news in the decade that followed. There were three reports of periodic damage caused by continuing ash falls from Mount Tavurvur to food gardens and marine resources on and around the small islands of East New Britain. There was one report of scientists continuing to assess the impact of the 1998 tsunami, and another report of the efforts of the local Catholic parish to reconstruct the livelihoods of villagers affected by it.

The after-effects of the 1998 tsunami included a heightened level of anxiety about this kind of peril. In 2004, it was reported that scientists were undertaking a study of the tsunami caused by a massive eruption on Ritter Island, between the New Guinea mainland and the large island of New Britain, back in 1888. One month later, it was reported that residents of other islands in the Siassi island group were in a state of panic about the prospect of a recurrence of that event. Three years later, provincial government officials appeared to confirm their fears by reporting another tsunami caused by an eruption of Ritter Island, though it does not seem to have killed or injured anyone. When a really big tsunami hit the coast of Japan and destroyed the Fukushima nuclear facility in March 2011, community leaders from two different parts of the Bismarck Archipelago reported that the wave had reached the shores of their own small islands.

Extreme Weather Events

Most of the big waves that threaten small islands and other coastal communities have nothing to do with volcanic eruptions. One of the biggest stories in my collection, accounting for more than 20 items in the newspapers, was about another kind of big wave or high tide that caused extensive damage across the Bismarck Archipelago and the islands of the Bougainville region in December 2008. The government officials who were the primary source of this story blamed this event on La Niña. Representatives of the Red Cross estimated that 60,000 residents of coastal communities suffered food shortages as a result of the damage caused to gardens close to the shoreline. Similar events on a smaller scale were reported from the islands of the Fly Estuary in 2005, and from the atoll communities of Bougainville in both 2005 and 2006. These were either described as king tides or as storm surges.

Drought was the other type of extreme weather event that reportedly caused serious food and water shortages, but there was nothing comparable to the droughts of 1997 or 2015 during the period in question. There were less than 20 newspaper items about the effects of drought on small island communities during this period, and these appear to indicate that different provinces were affected at different moments in time: Milne Bay in 2005, 2009 and 2013; New Ireland and New Britain in 2008; New Ireland again in 2010; Manus and Bougainville in 2011. The islanders of Milne Bay were also reported to have suffered some of the effects of Cyclone Yasi, which caused a great deal of damage along the coast of Queensland in February 2011, but there was only one newspaper item on this topic. Milne Bay is commonly regarded as the only province in PNG that is especially vulnerable to this type of natural hazard (McAlpine and Keig 1983).

Sinking Islands

Very few of the newspaper stories in my collection made any explicit connection between climate change and the incidence of extreme weather events. On the other hand, there were roughly 25 items that treated rising sea levels as an observable effect of global warming and an existential threat to island livelihoods. Three of these items were editorials, all published in 2007, which had a specific focus on the Carteret Islands in Bougainville. These appear to have been linked to a feature article on the plight of this 'sinking paradise' that was published in February 2007, and this in turn was associated with the spotlight cast by the cameras of a British television news crew. Other contributors to the sinking island narrative over the period between 2004 and 2013 included international conservation organisations, scientists from the University of PNG, provincial government officials and local community leaders. While the atoll formations of Bougainville were mentioned in more than ten of the news items, others reported the same issue from various parts of the Bismarck Archipelago. One article published in 2012 claimed that a small island in Manus Province (Ahus) was actually sinking faster than the Carteret Islands (Tiwari 2012).

The number of items in the story about rising sea levels was matched by an equivalent number of items in the story about one particular response to this affliction, and this second story was almost entirely devoted to the problems confronted by the Carteret Islanders. Most of the items in this second story were devoted to a contest between the Autonomous Bougainville Government and a community-based organisation, the Tulele

Peisa Association, each of which had its own plan to resettle the islanders, either on Buka or on the main island of Bougainville. As in the case of the Manam Island story, the focus therefore shifted from the question of why the islanders needed to be resettled to the question of how and where they should be resettled, so resettlement plans became a problem in their own right. In 2008, it was reported that the chief architect of the government plan, who was himself a Carteret Islander, had arranged the deportation of a German film crew because they were taking sides with the opposition. But this contest seems to have reached a kind of stalemate, since very few of the islanders were resettled anywhere (Connell 2016).

Degradation or Conservation of Marine Resources

Even if the story of resettlement is treated as an extension of the sinking island narrative, news items that clearly belong to the discourse of climate change were almost matched by those that dealt with the various ways in which the residents of small islands were suffering at the hands of foreigners in boats. That is partly because of the appearance of one big story in 2005, containing 12 items that dealt with the detention and subsequent release of some Chinese fishing boats in Milne Bay Province, although a small island was only named in the first of these items. There were several small stories about the incursions of Indonesian fishing boats, mostly in areas close to the Indonesian border, or the incursion of other foreign fishing boats whose origins were not specified. All these incursions became newsworthy because they were deemed to be illegal. Even if the foreigners were not fishing, their boats were sometimes accused of degrading the marine environment in other ways.

While this narrative portrayed the islanders as innocent and outraged victims of unscrupulous foreigners, there were occasional reports of malpractice on the part of the islanders themselves. In 2007, there were three articles about islanders from the Fly Estuary getting themselves arrested for illegally fishing in Australian waters. In 2005 and 2008 there were reports of islanders being killed or injured when using dynamite to bomb the fish on nearby coral reefs. And in 2013, a local scientist accused the residents of one urban island of polluting the local lagoon with their disposal of solid waste (Anon 2013).

There was a much larger number of items—about 70 altogether—that were concerned with the relationship between local fishers and the traders to whom they sold the products of their labour or a number of other external actors who were seeking to improve the productivity or sustainability of their harvesting practices. These other actors included the National Fisheries Authority and a range of non-government organisations supported by international donor agencies. Since these other actors were the sources of most of the stories that got published, the traders were generally treated as a menace, but the narrative still allowed that they had some of the islanders on their side. This tension came to a head in 2009, when the National Fisheries Authority imposed a ban on the harvest and sale of sea cucumbers (or bêche-de-mer) because of concerns about the sustainability of this fishery. This prompted a number of complaints about the negative impact of the ban on island livelihoods in several parts of the country. However, most of the stories in this general category made no mention of the traders engaged in the export of locally harvested products. They were only concerned with the progress that was being made in the establishment of marine protected areas, the implementation of novel mariculture projects, the installation of fish aggregating devices, or the conduct of scientific research on marine ecosystems and organisms. The foreigners who were investing in marine conservation projects obviously issued a lot more press releases than the foreigners engaged in the purchase and export of what the islanders were extracting from these ecosystems, but the publicity generated in this way did create an impression that the islanders needed a good deal of external support in order to deal with their problems.

What a Difference a Mine Makes

One small island accounted for 150 news items, which is more than 25 per cent of all the items in my collection. This is Simberi, which is a small island, though not a very small island, in the Tabar group of islands in New Ireland Province. Simberi became newsworthy during the period from 2004 to 2013 because it was the only small island that hosted a major mining operation during that period. There was a much bigger mining operation on a somewhat bigger island in New Ireland Province, which is the main island in the Lihir island group. If I had chosen to define this as a small island, rather than a medium-sized island, then the Lihir mining project would account for more than the 550 items in my collection of stories about newsworthy things happening on small islands. That is partly because the operators of major resource projects produce even more press releases than organisations engaged in nature conservation or environmental protection. And the government officials charged with the regulation of such projects, or members of local communities directly affected by them, as well as environmentalists, are all able to join in the chorus.

Simberi did rate a mention as one of the islands affected by the tidal wave reported from many parts of the country in December 2008, but if it were not for the mining operation, this might well have been the only occasion on which it made the news. Before that, most of the stories about the island followed the logic of the mining project cycle. The first story, containing 16 items, began in May 2004 and dealt with the feasibility studies being conducted by the project proponent, Allied Gold Ltd. The second story, containing 25 items, began in April 2005, and dealt with the construction of the mine following the grant of the relevant development licences by the national government. The third story, containing 5 items, began in February 2008 and dealt with the start of mining operations. The mining company was the primary source of all these news items, so this was all good news.

But then came some problems. The tidal wave at the end of 2008 does not seem to have affected the mining operation, but one year later the government's mining inspectors suspended the operation for a few days because of its failure to comply with some regulations, and the mining company had to issue a special set of press releases to let the public know that this problem had been solved and mining operations had been resumed. That story accounted for 16 items. But soon afterwards, in January 2010, a new story emerged. This one was not derived from the press releases issued by government officials or company managers. It was based on complaints by local landowners about the employment of Fijian security guards. That story occupied 14 items in the national newspapers. Then came several months of silence that was broken in February 2011 by the beginning of a story about the implementation of the benefit-sharing agreement that the company had made with representatives of the local landowners before the development licences were granted. This story resurfaced in January 2012, and then again in April 2013, when the benefit-sharing agreement was being reviewed. This story accounted for 28 news items over a period of three years.

Shortly after the beginning of the story about the benefit-sharing agreement, in February 2011, there was another story about the suspension and resumption of the mining operation. This time, it was suspended by officials from the Department of Environment and Conservation, not the mine inspectors from the Mineral Resources Authority, and the reason was a leak of toxic chemicals from the mine processing plant. This was a source of additional concern to the local landowners, whose representatives complained about the possibility of a second discharge of toxic material in March 2012, even though mining operations were not suspended at this juncture. Altogether, this story about the pollution of the natural environment accounted for 27 items in the newspapers, so it could be bracketed together with stories about other sources of environmental pollution.

Despite these distractions, mining company managers continued their standard practice of issuing periodic press releases about the continuation of the mining operation, some of which found their way into the newspapers. In July 2012, it was reported that the operation had been acquired by another mining company, St Barbara Ltd, but it appears to have run into some fresh difficulties by the end of 2013, when it was reported that a number of the mineworkers had been laid off.

By and large, these stories about Simberi were not stories about an island; they were stories about a mine. These stories also exemplify the difficulty of deciding whether a newsworthy phenomenon—in this case a major mining operation—constitutes a problem or a solution. Advocates of the mining industry would argue that it constitutes a solution to the problem of rural poverty, while environmentalists would argue that it constitutes a threat to the natural environment. The islanders themselves—or at least the customary owners of the mine lease areas—are obliged to weigh up the costs and benefits on their own account and make a noise when they are not satisfied with the result.

So What is Special about Small Islands?

If the development of a major resource project on a small island is a contingency that has nothing to do with the size of the island, it is likewise hard to see how the incidence of tectonic disruptions and extreme weather events, or the tension between the degradation and conservation of marine resources, is more of an issue for small island communities than it is for other coastal communities, including those on the mainland of New Guinea. Even the sinking island narrative, which does focus on small islands, has included some stories about the actual or prospective effects of sea level rise on these other coastal communities. While this narrative has highlighted the afflictions of communities based on atoll formations, like the Carteret Islands, it must be borne in mind that they account for a very small proportion of PNG's small island communities. So we must wonder whether mediated narratives of pressure have anything very specific to say about the problems that their members face.

It is worth noting here that there was only one newspaper article throughout the whole of this period that pointed to population pressure or land shortage as a distinctive type of affliction. This was an article reporting the views of a local government councillor from the Duke of York Islands in East New Britain Province, who thought that some of his people would need to be resettled on the main island of New Britain in order to resolve the problem (Anon 2007a). Limits on the supply of terrestrial ecosystem services to the residents of densely populated small islands only become newsworthy when the supply is suddenly reduced by some sort of natural disaster.

More common were articles that dealt with the problems arising from the absence of frequent or reliable means of transport linking small islands to bigger islands or the mainland. There were almost 20 items in this general category, but the stories had different angles. Five of the items reported the loss of banana boats at sea. One boat capsized in bad weather while the others were cast adrift when their outboard motors broke down. In three cases it was reported that the missing boats and their occupants had turned up on some distant shore. There were ten items that identified the problems that islanders faced as a result of the absence of more reliable and frequent means of transport, or the promises made by politicians to remedy this affliction, or the fulfilment of such promises. These included one article in 2011 about the launch of a new vessel, the MV *Bougainville Atolls*, whose name is sufficient to designate its purpose. But before that, in 2007, a government official reported that a new boat delivered to the Carteret Island community had become the source of a dispute between the islanders, rather like the plans for their resettlement.

Problems and Solutions

If a large-scale mine can be regarded as a problem for the natural environment or a solution to the problem of rural poverty, the same ambiguity attends other forms of 'development' that are less newsworthy because they have a smaller impact on ecosystem services or local livelihoods. The occasional visit of a boatload of tourists or individual buyers of bêche-de-mer would both exemplify this ambiguity. However, smaller stories tend to be less ambiguous because they are more likely to derive from a single source with a specific point of view. So the tourists count as good news when the source is the company that brings them ashore, while the traders count as bad news when the source is an environmental organisation intent on the establishment of marine protected areas. Stories about the mine on the island of Simberi were a mixture of good news and bad news because the voice of the mining company had to contend with the voices of community leaders or government officials who were occasionally upset about the risks and impacts of the operation.

Some stories join specific problems with specific solutions in a fairly simple and obvious way. Extreme weather events prompt the delivery of emergency relief supplies. Community leaders and government officials report the problem, government officials announce the solution, and community leaders sometimes complain about its implementation. But activities that seem at first sight to be straightforward solutions to a specific problem can also become a problem in their own right when they give rise to disputes about their implementation. Resettlement plans are a clear case in point, as exemplified by stories about the resettlement of the Manam and Carteret islanders.

Resettlement also counts as a solution (if it is a solution) to more than one of the problems that islanders face—in this case as a rapid response to a sudden volcanic eruption or a slower response to gradually rising sea levels. Nor is resettlement the only possible solution to the problem identified in the sinking island narrative, despite the weight accorded to this solution in the case of the Carterets. There were also a few stories about islanders being trained or encouraged to plant mangroves or build sea walls to deal with this problem or the problem of occasional storm surges. These stories had a focus on small islands in Manus and New Ireland provinces, not the atoll formations of Bougainville, because these happened to be the islands selected for donor-funded experiments in climate change adaptation.

The newsworthiness of the sinking island narrative concealed the extent to which climate change was understood to be the source of the problems to which various solutions were being offered. The connection was sometimes less evident in the stories themselves than in the sources from which they were derived. For example, most of the 25 items about efforts to improve food, water or energy security, or the sustainability of island livelihoods, were sourced from press releases produced by donor-funded projects whose titles revealed their rationales, like the Pacific-Australia Climate Change Science and Adaptation Planning Program, funded by the Australian Government, or the Atoll Research and Development Project, funded by the European Commission. Roughly half of the 20 items about the conduct of scientific research also indicated that climate change was the source of the problem being studied, but these stories rarely suggested any particular solution to the problem. This is not surprising when one considers that the problem being studied might be a physical process like ocean acidification or saltwater intrusion.

Stories about the conduct of scientific research, which were typically sourced from the scientists themselves, were comparable to a few stories about the activities of foreign film crews. What was newsworthy was not so much the problem in itself but the process by which it was being investigated. All of the seven items about the arrival of foreign film crews were concerned with the atoll formations of Bougainville, and six of these were clearly contributions to the sinking island narrative. The seventh item was about a film crew that visited the Mortlock Islands, presumably intending to add to this body of publicity, but found that the islanders had been distracted by the arrival of a foreign fishing vessel which they had impounded. Six of the seven items were clearly sourced from the film crews or their sponsors, so counted as acts of self-promotion. The remaining item was the one that reported the deportation of a German film crew whose contribution to the sinking island narrative had upset the Carteret Islander who was a senior public servant in the Autonomous Bougainville Government. In a sense, these were all local stories about global stories, or stories about the way that local problems and solutions were being represented to a global audience. As in the case of the resettlement plans that the German film crew was investigating, it is not clear whether this form of mediation counts as part of a solution or part of a problem.

One Outlandish Story

I shall now focus my attention on a group of eight news items, out of the 555 that I collected over the period between 2004 and 2013, that appear to constitute a truly outlandish story, in the sense that I could think of no way to group it together with other stories, about other islands, that make immediate sense as stories about specific forms of peril or pressure, or about the manner in which people dealt with them. The story about the visit of a luxury cruise liner to a number of small islands in 2004 was somewhat outlandish in the sense that it did not seem to have much to do with any specific problem or any specific solution to that problem. However, it still had something in common with other stories about the arrival of tourists or scientists or journalists who had some particular interest in the islands or their occupants, since the stories were primarily about the nature of that interest. The Emirau story is outlandish because the island's problems were not what they first appeared to be, and that in turn explains the peculiarity of the solutions that were being canvassed and debated.

Emirau is a small, though not very small, island, with an area of approximately 34 km². It belongs to the St Matthias group of islands in New Ireland Province, and is approximately 130 km northwest of Kavieng, the provincial capital. It is not too far (less than 30 km) from the medium-sized island of Mussau, which is the biggest island in the group, and since Mussau boasts a health centre that is (probably) capable of dealing with a medical emergency, Emirau is not as isolated as its distance from Kavieng might suggest. The 2000 census recorded a rural village population of 500, distributed between six villages located around the coastline, all part of a single council ward in the Murat (or St Matthias) LLG area. In 1979, about 24 per cent of the total village population was apparently absent from the island, and many of these people were likely to have been resident in Kavieng. Emigration from the island appears to have continued between 1980 and 2000, since the resident population grew by less than 12 per cent over that period. But none of these numbers make Emirau seem in any way unusual.

Emirau was mentioned in despatches as one of the many islands in New Ireland and New Britain that were afflicted by drought-induced food shortages in April 2008 (Vuvu 2008), and then as one of a larger group of islands across the Bismarck Archipelago and the Bougainville region that suffered the effects of the king tides or storm surges recorded in December of the same year (Eroro 2008; Nicholas 2008). But here again there is nothing that sets Emirau apart from other small islands in the same boat, nor do these extreme weather events figure in the story that does make Emirau stand out from the rest of the small islands that feature in my collection of newspaper stories.

Great Expectations

The first item in the group of eight that constitute this outlandish story was a letter to the editor, not an article written by a journalist or one based on a press release by some influential person or organisation. The letter was entitled 'Cult-Like Storm Brewing on Island'. It was published in December 2004, long before the island was visited by the extreme weather events of 2008. The author of the letter described the island as 'a beautiful place filled with peace-loving Seventh-Day Adventist people', and then went on to deplore the 'cult-like storm' as the effect of 'developmental schemes designed and promised by certain "prominent" persons without ever eventuating'. These schemes were said to appeal to the islanders because of their disengagement from the cash economy, which was in turn due to their 'isolation and lack of safe and consistent sea transport'.

> This point is evident in their current plan for a massive relocation of the islands' six villages for a major fish project with an international airport for direct flights to Japanese markets and space for five-star hotels on the tiny island. The current plan is said to be contributing a similar dream-state or cultic effect to that which was experienced in early 1994 and late 1995, where the entire island populace was mobilised in preparation for a rocket-launching space project that never occurred.
>
> (Siko 2004)

So this looked like the beginning of a story about a familiar set of problems arising from the absence of frequent or reliable means of transport linking small islands to bigger islands or the mainland. Nothing unusual about that. But what made this story stand out from those with a similar theme was the 'cult-like storm' that was supposedly associated, like an extreme cultural weather event, with the solutions being promoted by 'prominent' but unidentified people. Although the idea of a 'cargo cult' does occasionally surface in the pages of the national newspapers, when rural villagers are said to be doing strange things, or subscribing to strange ideas, there is no other news item in my collection that makes reference to 'cult-like' thoughts or actions being adopted in response to any of the problems encountered by the residents of small islands.

The next episode in this story was another letter, presumably also written by a member of the Emirau community, albeit resident in the national capital, which was published in September 2005 (Aute 2005). This letter identified the most 'prominent' member of the same community as the principal agent in the 'developmental scheme' that had failed to impress the author of the previous letter. This person was Ben Micah, former MP for the Kavieng Open electorate and long-time supporter of former prime minister and fellow New Irelander, Sir Julius Chan. Micah was said to have been negotiating a deal with the national government and an Australian called Edward Carr to invest millions of kina in a combination of 'fishing, redevelopment of old World War II airfields on the island, tourism and communication development'. A certain amount of seed capital might have been required for this venture, since the correspondent also recorded that

Micah had organised a 'fundraising' event in Port Moresby in March 2005, and that is where he is said to have announced the imminent finalisation of the whole deal. The author of the letter did not think there was anything preposterous or unrealistic about the development proposal, and rather thought that the islanders were quite 'excited' about it, but he was also keen for further news of its progress.

Readers of the national newspapers would have to wait more than two years for such news to make an appearance. In December 2007, two articles were published, both apparently based on press releases emanating from the office of Sur Julius Chan, who had re-entered the national parliament as governor of New Ireland Province in July of that year.

In the first article, the governor announced that his provincial government had set its seal of approval on what was now described as 'an integrated business strategy and practices which project immense monetary gains', not only for the people of Emirau, but also for the Murat LLG area, the rest of the province, and even the rest of PNG (Anon 2007b). But Emirau was clearly the main focus of attention. Indeed, this was said to be a 'home-grown project' initiated by the islanders themselves, and one that was now coming to fruition as a result of several years of negotiation that had 'settled questions of land mobilisation and registration'. These negotiations had apparently involved the '29 clans' that contained members of the island community, represented by their 'parent company', Emirau Trust Ltd, under the chairmanship of a former pastor called Stephen Wilson, 'with help from overseas consultants'. If the reader was still wondering what was so special about this particular island, a sort of answer was provided at the end of the article, where it was noted that Emirau had been 'chosen by the Americans as a strategic defence base' during the Second World War, and so, by implication, foreign investors would now make a comparable choice, and the islanders could well understand why they would do so.

The second article provided some additional information about the nature of the foreign investment that was now being anticipated. It would apparently involve 'the development of an "all weather" international deep-sea port for shore based fishing operations, general freight and cruise ships'; it would constitute 'an integrated free trade zone, fishing, tourism and transport hub project'; and so the island would become 'the new distribution point for all inbound and outgoing cargo from industrialised countries of the north and the southern regions through PNG in the shortest and [most] cost effective manner' (Anon 2007c).

Ben Micah's name was not mentioned in either of the articles published in 2007. However, in September 2008 he issued a press release of his own in which he announced that he had submitted a proposal for what was now described as 'Project Emirau' to the Department of Commerce and Industry on behalf of Emirau Trust. The idea now was that Emirau would become an especially special free trade and export processing zone within a 'special economic zone' that would encompass the whole of New Ireland Province. The article based on this press release also advised the reader that the islanders 'through their 29 incorporated land groups, [had] formally assigned all their customary land rights and claims of the island to the Emirau Trust on Jan 10, 2005', so this entity was now 'the leaseholder and sole landlord of the Island and the only authorised entity to deal with the Government and private investors regarding developments' (Philip 2008).

Another Narrative Landscape

Three more years passed by before the next episode in the story, and this episode served to confirm that Emirau had been absorbed into a narrative landscape that had nothing to do with small islands. Three articles were published in the national newspapers in rapid succession in September and October 2011. The first article reported that the island had become the private property of Emirau Trust, and that this entity was owned by a single individual (Paniu 2011). The second article reported that the Trust had transferred its new-found property rights to an Australian businessman, and that some of the islanders were complaining that they had never consented to the loss of their customary land rights (Anon 2011a). The third article reported that Ben Micah was preparing to launch defamation proceedings against the newspaper for suggesting that any such act of alienation had taken place (Anon 2011b).

All three articles were derived from the preliminary hearings of a commission of inquiry that had been set up to investigate what PNG's Land Act calls 'special agricultural and business leases' (SABLs). Readers of the *National Gazette*, which is the government's own official newspaper, would have known that the island of Emirau had been subject to one of these leases in December 2006, and that the lease had been issued to Emirau Trust for a period of 99 years (GPNG 2006). I was made aware of this fact by one avid reader of the *National Gazette* who had become concerned about the possibility that these leases over customary land were being granted without the free, prior and informed consent of the customary landowners. I had already begun to collect newspaper stories about these leases, distinct from my collection of stories about small islands, well before the commission of inquiry was established, because I was one of the participants in the campaign that led to its establishment (Filer 2017).

The 3,384 hectares of land covered by the Emirau lease were just a drop in the ocean of more than 5 million hectares of customary land that had been covered by such leases by 2011. The leases of most concern covered much larger areas of land and were associated with proposals to fund large-scale agricultural projects through the wholesale clearance of native forests (Gabriel et al. 2017). Small islands were too small to be of any interest to the proponents of these so-called agro-forestry projects. While many SABLs had been issued for other purposes, and many covered fairly small areas of customary land, hardly any had been issued over small islands for any purpose whatsoever. So the Emirau lease is an oddity among the 85 leases investigated by the commission of inquiry, just as Emirau is an oddity in my collection of newspaper stories about the problems of living on a small island.

The establishment of a commission of inquiry meant that items published in the national newspapers became a secondary and somewhat superficial source of information about what had happened on this island, which was now quite clearly a small story of island alienation in a nationwide narrative landscape. The newspapers did not report the detailed hearings into the Emirau case that were conducted by Commissioner Nicholas Mirou at the Fisheries College in Kavieng on 26 and 28 October 2011 (Mirou 2011a, 2011b). From the transcripts of these hearings, we can trace the origin of the story back to 1994, when the president of a body called the Emirau Landowners Association accused the president of a body called the Emirau Development Corporation of trying to claim ownership of the whole island when there was talk of building a satellite launching station. Ten years later, in November 2004, Ben Micah, in his purported capacity as the 'Chief of Emirau Island' signed a memorandum of understanding with Edward Car [sic] of Kangaroo Ground, a small town near Melbourne. This agreement apparently gave Mr Car exclusive rights to all the fish harvested within a 12-mile radius of Emirau.[7] He would also be given the right to build an international airport capable of accommodating the largest passenger aircraft then in operation and charge them landing and take-off

---

7   Under PNG law, the only entity that could grant such a licence is the National Fisheries Authority. The members of coastal communities do not have customary fishing rights that could be alienated in this way.

fees. It was also agreed that Mr Micah would facilitate the creation of an island economy in which the US dollar would replace the PNG kina as the official currency. For his part, Mr Car promised to pay a monthly rent to Emirau Trust, to build a wharf, as well as an airport, and a waste recycling plant, as well as a fish processing facility. According to counsel assisting the inquiry, the waste recycling plant was 'presumably for all the metal and other junk left [behind] by the United States Army during World War II' (Mirou 2011a: 8).

Emirau Trust was then incorporated, with a single shareholder and director, one Elvee Judy Mave, with the stated intention of sub-leasing the whole island to Mr Car for a period of 99 years. It was at this juncture, in December 2004, that one of the dissident landowners wrote his letter about the 'cult-like storm' brewing on the island (Siko 2004). Nevertheless, in April 2005, the *National Gazette* published a notice to the effect that the 29 land groups on the island had been formally incorporated and registered (GPNG 2005). The decision of the land group executives to alienate all their customary land rights to the state was certified by the acting provincial administrator in August of that year, and that is what made it possible for the state to issue an SABL to Emirau Trust at the end of 2006.

Executives of the Emirau Landowners Association continued to challenge the process of alienation and to deny Ben Micah's claim to be their 'chief' or their 'principal landowner'. In November 2006, a government official convened a meeting in the provincial capital in an effort to reconcile the project's supporters and opponents, but the meeting turned into a fight and several of the participants ended up in hospital (Mirou 2011a: 10).[8] However, those who appeared before the commission of inquiry conceded that most of the people still living on the island had consented to the new deal 'because they heard of big things coming … big money coming, tangible projects coming' (ibid.: 41). And yet the route by which they would arrive was still uncertain. A representative of Emirau Trust told the inquiry that the agreement with Edward Car had already lapsed by the end of 2010 and had been replaced by an agreement with another potential developer of the 'special economic zone' after the New Ireland Provincial Government had contributed a sum of 300,000 kina that was spent on legal fees and travel costs (ibid.: 26–7).[9]

8   It seems that the opponents later went to court to challenge the validity of the lease agreements (Anon 2011b). However, these legal proceedings were not mentioned in the hearings conducted in Kavieng.
9   At that time one PNG kina was worth about 42 US cents.

In his final report, Commissioner Mirou recommended that the SABL be allowed to stand on condition that Ms Mave be replaced by 29 land group executives as shareholders and directors of Emirau Trust and that further efforts should be made to resolve the dispute between the two landowner factions (Mirou 2013: 270–1). His failure to recommend the cancellation of the lease appears to have been based on his recognition that a majority of the islanders had agreed to it.

The Space of Dreams

Edward Car did not make an appearance before the commission of inquiry, but he did create a website. A denuded version of this website still exists at time of writing.[10] It shows Mr Car surrounded by happy smiling children, presumably children of Emirau Island, and includes a promise that he is still looking for ways to fund his project.

A previous version of this website, still available at the time of the inquiry, was far more interesting. In that version, Car said that he first visited the island in 2003 in his capacity as 'the Leader of a Value Based Expedition run by WIND AUSTRALIA visiting the most distant point on the map that had land in the Pacific Ocean in the most remote corner of PNG'. Although he had some very general ideas about the island's actual and potential place in the global capitalist system, Car represented Project Emirau as a joint venture between Emirau Trust and a company called Wind Trader Ltd (perhaps his own family company) to develop something called the 'Emirau Mother's Haus'.

The ideas behind this entity are set forth on another website.[11] It is described as 'a contemporary modern living air conditioned high tech Home Office designed around the mother and her roll [sic] in the community', and as 'a self sufficient building that provides its own Permaculture Food, Rain Water, Green Electricity, Grey and Black water treatment, Methane for cooking and cooling and Telecommunications and Internet'. The cost of the building itself was estimated to be quite modest, between 30,000 and 40,000 US dollars, but the cost of relocating all the islanders to a cluster of new buildings around it would be much higher—between 4 and 5 million US dollars. Car claimed that a bank loan used to finance the development of this new community infrastructure could be paid off within ten years.

---

10   Archived at web.archive.org/web/20221208153227/http://emirau.asia/index.html
11   Archived at web.archive.org/web/20220316091826/http://emirau.asia/e_house.html

The cost of the new housing would be covered by the sale of 'Export Organic Food and Water, Export Wild Fish, Boutique Accommodation and Electricity', while the cost of the entire project could be covered by a 'Building Material Production Plant' selling building materials 'to the rest of Melanesia'. The sale of other items like 'Tuna Sashimi Fish' and 'Medical Tourism' would simply add to the overall profitability of the scheme. This would indeed have been a very special kind of special economic zone if it had ever been created. But time had already passed it by when Commissioner Mirou conducted his hearings in 2011.

The Wheel of Fortune

Another potential witness who did not make an appearance before the commission of inquiry was Ben Micah. The hearings do not seem to have made much of a dent in Mr Micah's political capital, since he was re-elected to the national parliament as the MP representing the Kavieng Open electorate in July 2012. He was then appointed as the minister for public enterprise in the government of Peter O'Neill—a position that he held for the next four years.[12] In this capacity he could work with his patron, Sir Julius Chan, to pursue a new version of a much older development plan.

In January 2015, a press release from the provincial governor's office announced that his government was holding talks with an 'international consortium' to develop a satellite launching facility on Emirau Island 'within a couple of years'. This was represented as the resurrection of a scheme originally proposed by the government of the Soviet Union in the 1980s, which might well have become a reality if the Soviet Union had not disintegrated. Emirau was said to be the obvious and natural choice for a facility that was bound to attract the interest of 'several international satellite launch companies' because

> the Earth is actually moving faster at the equator than at any other point, [so] it reduces the amount of fuel necessary. So if you have to lift less fuel you can lift more equipment and materials, which makes it much more cost effective.
>
> (Anon 2015)

---

12 Micah lost his seat at the national elections held in 2017. He died in March 2022.

While this might sound like a rather more plausible proposal than those put forward by Mr Car, nothing more has been heard of it. So the 'rocket-launching space project that never occurred' (Siko 2004) is still awaiting its moment of realisation.

Discussion

The pressures to which Emirau Island has been subject are barely visible through a pair of cartographic and demographic spectacles, nor do they seem to be typical of the sorts of problems that confront the residents of many small islands in PNG. Instead, they seem to have arisen from specific moments of disruption that have a long and peculiar history.

The first of these moments was the partial alienation of the island for the development of a copra plantation during the German colonial period, an experience that was indeed shared with many other small islands in the Bismarck Archipelago (Nevermann 1933). However, the plantation land was apparently restored to its customary owners in the early 1970s (Mirou 2011a), and this was also a common experience.

The second moment of disruption was the sudden and dramatic conversion of the islanders by Seventh-Day Adventist missionaries in 1931. This experience was shared with the rest of the St Matthias group of islands (Dixon 1981). According to local legend, the missionaries were the first foreigners to induce what the first newspaper correspondent described as a 'dream-state or cultic effect' among the islanders (Siko 2004). The legend has it that the islanders were so enchanted by the sound of hymns wafting from the missionaries' boat that they instantly forsook their pagan customs and became what Joel Robbins (2004) calls 'morally tormented sinners'. Furthermore, their radical conversion brought about an equally radical change in their management of local ecosystem services because they killed all their pigs and stopped eating all the wild abominations of Leviticus (Christin Kocher Schmid, personal communication, 1997).

But Emirau, unlike the other islands in the St Matthias group, was to face a much bigger disruption to its social–ecological system during the Second World War, when it hosted the American airbase that left the islanders with a long-term legacy of metal and concrete 'junk'. Once again, this was not an entirely unique experience. Other military airfields were constructed on Nissan Island, the main island in the Green Island atoll formation in

Bougainville, on the peri-urban island of Los Negros in Manus Province, and even on the miniscule island of Ponam, also in Manus Province.[13] However, Emirau does seem unique in the extent to which this wartime legacy encouraged the production of apocalyptic dreams that entail a whole new form of industrialisation, possibly because there was a very large amount of 'junk' to be contemplated, or else because the island had natural features, like a harbour, that made it seem like a suitable venue for one or more of the projects that were envisaged.

From this we might infer that each small island has its own singular moments of colonial history, which are more or less well documented, and that is why it makes no sense to say that they are 'all in the same boat'. But a distinction can still be made between moments of extreme disruption, which transformed the relationship between island communities and island ecosystems in ways that were both profound and irreversible, and moments of temporary disturbance, or more gradual change, in which elements of the relationship could display the property now sometimes known as 'resilience'. And this distinction applies just as much to changes experienced in the post-colonial period or the pre-colonial past. If all these histories were better known, then small island communities could be classified by reference to their experience of pressures and perils that are not revealed by the 'facts' of cartography and demography. But in the absence of such an evidence base, the only things that can be classified are different forms of disturbance or disruption.

The volcanic eruption that took place on Manam Island in 1994 has proven to be a moment of extreme disruption, and the mine that arrived on Simberi Island in 2005 will most likely prove to be another one. Large-scale mining operations belong to a class of 'development projects' that include the copra plantations of the colonial period and the military airbases built during the Second World War, and might now include the construction of tourist resorts or the various projects conceived by Ben Micah and Edward Car. This kind of disruption could only occur as part of the process by which small islands have been selected as sites of industrial development or capital accumulation in one of its various forms. The disruptive effects of natural hazards, like volcanic eruptions or extreme weather events, have been felt

---

13 In 2000, the rural village population of Ponam was almost as large as that of Emirau, while the island itself was more than 30 times smaller, so the legacy of concrete 'junk' would make a much bigger dent in the amount of land available for village agriculture when the war was over (Carrier and Carrier 1989: 82).

for as long as vulnerable islands have been occupied by human populations. But this class of 'natural' disruptions is now in the process of being expanded to include the effects of anthropogenic climate change, so a small island may be vulnerable to the effects of industrialisation without being the site of any specific form of modern industry.

One of PNG's island communities clearly stands out from all the rest as the victim of climate change by virtue of the position it has come to occupy in both national and global versions of the sinking island narrative. That is the community of Kilinailau, more generally known to the outside world as the community that inhabits the Carteret Islands in the Autonomous Region of Bougainville. According to John Connell (2016), the rate of sea level rise in the Bougainville region is far too low to explain why this particular group of islands has attained its 'iconic' status as home to the world's first 'climate change refugees'.[14] Instead, he argues that the Kilinailau community has achieved this singular status because it has been subject to an entirely different set of material pressures and perils, and the islanders have embraced their new reputation because of their inability to deal with the extreme nature of these other problems. The most basic problem has been growing population pressure on a limited resource base, and this has only been accentuated by the failure of a number of possible solutions. The Carteret Islands have now been 'marginalised through the creation and reiteration of an emotional geography' in which another problem— the problem of climate change—has become a kind of distraction from the absence of any workable solution to the real problems that beset this community (Connell 2016: 7). Through the advent of a wave of international publicity about the plight of sinking atolls, 'a negative climate change discourse has been converted into both a romantic essentialism of place and a "weapon of the weak"' that constitutes 'a distraction from the necessity of achieving daily livelihoods' (ibid.: 2012).

There is no doubting the extent of the population pressure. The 2000 national census recorded a resident rural village population of 979, which was 25 per cent higher than it had been in 1980. At this rate, Connell reckons that the number of residents would have been about 1,200 in 2015. Since the surface area of the entire atoll formation is barely 70 hectares, this suggests that the Carteret Islands have the highest population density of any non-urban atoll formation in the Pacific Island region, let alone in

---

14   In the absence of any precise measurements, he estimates the rate to be somewhere between 3 and 5 mm per annum (Connell 2016: 7).

PNG. In the late colonial period, when the population was a good deal smaller, and no one was talking about climate change, there were plans to build coral seawalls to deal with the problem of coastal erosion and plans to resettle some of the islanders in order to deal with the bigger problem of population pressure (Mueller 1972; O'Collins 1990). But Connell is more concerned with the problematic nature of another solution, which consists of the voluntary migration of some of the islanders—especially adult men—to places where they can find some kind of paid employment and support the remainder of the island community by means of remittances in cash or in kind.

As we have seen, it is difficult to assemble a body of evidence that would make it possible to test a hypothesis about the relationship between rates of out-migration, let alone the volume of remittances, and the extent of population pressure on small island ecosystems. In this particular case, we do not even know how many people were counted as 'absentees' from the Kilinailau community in 1979 because none of the Bougainvillean atoll formations were included in the provincial data system (PDS). Connell reckons that 300 of the islanders were absent in 2015, and that number presumably includes the handful of households that had been 'resettled' on the main island of Bougainville. An absentee rate of 20 per cent is almost identical to the average recorded for 19 miniscule and remote islands in other parts of PNG that were included in the PDS in 1979 (see Table 2.6). However, he also reckons that as many as 40 per cent of adult male members of the Kilinailau community were absent from the islands in 1982, just as they had been in the 1960s (Connell 2016: 9).[15] Whatever the numbers may have been, Connell is probably correct in suggesting a fairly drastic reduction in the 1990s, when the incidence of civil conflict on the main island of Bougainville caused many of the absentees from the atoll communities to seek refuge at home, but there is no clear evidence on the extent to which this process has been reversed in the 20 years since the peace agreements came into effect, nor any particular reason to believe that the Carteret Islanders have found it especially difficult to find new ways to leave home again. So perhaps the root of their problem is still the sheer number who would have to do so in order to keep the islands afloat in an economic, rather than a physical, sense.

---

15 It is not clear how many of these men would have been counted as 'absentees' in 1979.

So what lessons can be drawn from a comparison of island communities whose members have very different histories of dealing with different forms of pressure or peril?

The first lesson would be that natural hazards and 'development projects' do not exhaust the range of historical events that can disrupt the relationship between island communities and island ecosystems. In this respect, the sudden and dramatic conversion of the Emirau community by Seventh-Day Adventist missionaries can be compared to the effect of Bougainville's civil conflict on the livelihoods of the Kilinailau community, or even, by Connell's account, to the conversion of the Kilinailau community by the missionaries of climate change. These are not examples of a third form of disruption; they simply reveal the limitations of the initial dichotomy.

The second lesson is that dramatic and newsworthy disruptions may or may not have long-lasting effects on the relationship between island communities and island ecosystems, and may have less of a long-term effect than what are sometimes called slow drivers of change, like population growth or rising sea levels. Extreme weather events, like king tides or storm surges, are commonly cited as examples of what Connell (2016: 7) calls 'short-term cyclical events' that have a greater immediate impact but less of a long-term impact than population pressure. Bougainville's decade of civil conflict might also count as a temporary disruption to the process by which atoll communities like Kilinailau have been more or less able to deal with this deeper but less newsworthy problem. The same could be said of development projects or resettlement plans that never come to fruition, but only seize the imagination of island communities or people who write stories about them.

The third lesson is that the difference between fast and slow drivers of change in the relationship between communities and ecosystems, or in the state of 'social–ecological systems', is not simply a difference between types of physical process whose relative speed can be measured by scientific methods. Slow drivers can be turned into faster ones, and real drivers into imaginary ones, by mediated narratives of pressure. This is essentially Connell's point about the sinking island narrative. But the point can be taken one step further. The newly 'iconic' status of the Kilinailau community within this narrative may well have more substantial and material effects on their social–ecological system because of the sheer volume of publicity that has encompassed the Carteret Islands at national, regional and global scales. In other words, the identity of this island community has been transformed in ways that do not simply conceal its material circumstances but change its place in the world. If Ben Micah's plans had come to fruition, the Emirau community would be subject to a similar kind of transformation. So the way that islanders or outsiders perceive the pressures and perils of existence on any particular island, or the possible solutions to these problems, do not simply constitute a set of predictable 'responses' to more or less rapid changes in their material circumstances, but can have an independent and less predictable effect on material relationships.

This leads us to the fourth and final lesson, which has to do with the nature of human agency in the fabrication of island livelihoods. While Connell recognises the potential significance of what he calls 'local agency' in the Carteret Island case, it is not clear how this has made a difference to the way that the islanders have dealt with the problems of population pressure and food security. On one hand, he notes that the islanders were traditionally able to alleviate these problems through the exchange of locally made shell money for foodstuffs produced on the medium-sized island of Buka (Mueller 1972). On the other hand, he observes that the problems have more recently been aggravated by what he calls 'accelerated human influence' on the local environment, including the substitution of cash crops for food crops, the destruction of mangrove ecosystems, and the over-exploitation of marine resources (Connell 2016: 8). Now this might be construed as a sign of failure in the adaptive or managerial capacity of local social institutions (Bourke and Betitis 2003). But the whole point of invoking the significance of human agency is to avoid the kind of ecological or geographical determinism in which the fate of each small island community is somehow predictable from measures of 'pressure', 'isolation' or 'vulnerability', without reference to history, culture and politics.

The way that islanders perceive and respond to externally generated pressures and perils is itself a function of the social relations or networks that constitute their communities. Aside from those that are internal to the island on which they live, these include relationships between neighbouring communities, whether or not their members inhabit small islands, relationships between 'alien intruders' and the inhabitants of islands to which they are attracted, and relationships between islanders who stay at home and those who venture abroad. These four types of social relationships can interact in many different ways, and are subject to their own distinctive forms of disruption or disturbance. Their interaction and disruption cannot simply be predicted from the 'facts' of geography or demography, nor from the specific nature of the pressures and perils to which communities are subject.

Traditional networks of exchange between small island communities and neighbouring communities have been substantially disturbed, if not thoroughly disrupted, by the devaluation of many of the things that used to be exchanged, like shell money, clay pots or the ingredients of sailing canoes (Hogbin 1935; Belshaw 1955; Schwartz 1963; Harding 1967; Lauer 1970; Egloff 1978; Macintyre and Young 1982; Berde 1983; Irwin 1985; Lutkehaus 1985; Macintyre and Allen 1990; Terrell and Welsch 1990; Gosden and Pavlides 1994; Allen and Gosden 1996). It is a moot point whether contemporary patterns of migration and remittance have a similar effect on the sustainability of island livelihoods (Carrier 1981; Hayes 1993; Pomponio 1993; Guy 1997; Christensen and Mertz 2010; Connell 2010; Rasmussen 2015). But relationships between islanders who stay at home and those who venture abroad are not reducible to numbers, nor are they isolated from relationships with 'alien intruders'. The Carteret Islands have not hosted a succession of international film crews because climate scientists thought they were sinking at an unusually rapid rate, nor because government officials worried about the problem of overpopulation. Their global notoriety reflects the local agency of Ursula Rakova, founder and leader of the Tulele Peisa Association (Rakova 2014), and her role can be compared with that of Ben Micah, the self-styled chief of Emirau, in his efforts to attract foreign investors to his own island community. Both these individuals belong to a small island diaspora whose agency does not simply consist in the supply of remittances to relatives at home, but also consists in the supply of stories about the pressures and perils to which their communities are subject, and hence about the possible solutions to their problems. And these stories have their own effects.

Conclusion

So is it possible to arrive at a classification of small island communities by reference to different forms of marginality or marginalisation without lapsing into the simplifications of the pressure–state–response model? Or can they be situated at different points in a national or global process of uneven development that makes some of them more peripheral than others, regardless of their relative physical isolation or the extent of population pressure on a limited bundle of natural resources?

Connell's definition of marginalisation is clearly at odds with the one more commonly adopted in political ecology (Blaikie and Brookfield 1987). But that is partly because the conventional definition does not seem to fit this case. Few of PNG's small island communities appear to be marginalised in the sense of being blamed for a process of environmental degradation that is actually the result of material dispossession or exploitation by foreign investors or ruling elites. Most of the dispossession and exploitation took place in the early colonial period, more than a century ago, and traces of this colonial legacy are now hard to detect. A small number of communities have been affected by the development of large-scale mining projects, like the one on the island of Simberi, but the islanders are not averse to this kind of development, so long as the benefits outweigh the costs, and the moment of their marginalisation is therefore postponed to the time when the mine eventually closes down (Filer et al. 2016). A larger number of communities have been affected by the incursion of foreign fishing vessels, but the islanders have been able to counter this kind of encroachment with the backing of their own government.

Connell's definition of marginalisation is one that places a lot more emphasis on the effect of discursive practices and social imaginaries, which can also be construed as a form of dispossession and exploitation (West 2016). But whose practices and imaginations are at work here? The Kilinailau community is quite unique in the depth of its experience of foreign film crews looking for evidence of people's response to rising sea levels. A few other communities have been subject to a smaller amount of attention from the same source, while a somewhat larger number have attracted the gaze of foreign tourists seeking the romantic essentialism of an island paradise. Although these can be construed as distractions from the material problems of their existence, there is hardly any evidence to suggest that the islanders resent the attention, and rather more to suggest that they welcome it. Indeed, these imaginaries might have less of a deleterious effect than the attention paid by scientists who want to help communities find 'real' solutions to their 'real' problems.

A number of small island communities have been the subjects or targets of applied research projects funded by international aid agencies. These projects are typically framed by a global discourse that links the problem of climate change adaptation to the problems of food security, disaster risk reduction and the conservation of marine biodiversity values. The research typically involves a combination of surveys and workshops with members of a sample of coastal communities, including small island communities, with the aim of finding solutions that can be 'scaled up' or 'scaled out' to a wider collection

SMALL ISLANDS IN PERIL?

of communities that are assumed to have the same problems (Cinner et al. 2005; Kurika et al. 2007; Butler et al. 2014; Narayan et al. 2015; Maina et al. 2016; Purdy et al. 2017; Hair et al. 2019). Studies of this kind may be criticised for their adherence to the pressure–state–response model, even when they find that different communities have different responses to the same pressures, and therefore have to concede that common solutions to the same problems are not within reach (Butler et al. 2020). But their discursive practices are rather more problematic, and potentially deleterious, because projects like this can only last for a limited period of time. The budgets run out, the researchers retire from the field, and someone else is then responsible for implementing their recommendations. If government agencies are unable to do so, as is often the case, the communities are left to their own devices. If those devices are not up to the mark, then 'community-based resource management' becomes a recipe for frustration and demoralisation.

If Connell is correct in his assertion that migration is the only long-term solution to the real problems of the Carteret Islanders, this does not yield a general conclusion about the relationship between migration and marginality. Instead, it raises a different question about the classification of small island communities. Are these communities more or less peripheral or marginal if a certain number of their members are unable to leave their islands when they would rather move elsewhere, or if a certain number are obliged to leave their islands when they would rather stay put? Now this can be taken as a question about the feasibility of resettlement plans in response to rising sea levels or volcanic eruptions (Lipset 2013; Connell and Lutkehaus 2017; Gharbaoui and Blocher 2017; Luetz and Havea 2018). But that is only one aspect of the question. The factors that influence the capacity and desire of individuals or families to move back and forth between small islands and other places are not reducible to the pressures and perils of living on a small island as opposed to some alternative kind of place. So a classification that is based on changing patterns of migration would have to be based on a different set of considerations. And that is the point at which a closer reading of small island ethnographies would seem to be justified.

References

Allen, J. and C. Gosden, 1996. 'Spheres of Interaction and Integration: Modelling the Culture History of the Bismarck Archipelago.' In J. Davidson, G. Irwin, F. Leach and others (eds), *Oceanic Culture History: Essays in Honour of Roger Green*: Special issue of *New Zealand Journal of Archaeology*.

Anon, 2004. 'Luxury Liner to Berth at Humble Siassi Islands.' *Post-Courier*, 12 May.

——, 2007a. 'Duke of York Islanders Face Crisis.' *Post-Courier*, 28 May 2007.

——, 2007b. 'Chan Eyes Riches.' *Post-Courier*, 12 December 2007.

——, 2007c. 'Nod for NIP Deep Sea Port.' *Post-Courier*, 20 December 2007.

——, 2011a. 'Australian Owns Emirau Island.' *Post-Courier*, 30 September 2011.

——, 2011b. 'Micah Denies He Has Sold Emirau Island.' *Post-Courier*, 3 October 2011.

——, 2013. 'Lagoon Worries Expert.' *The National*, 31 December 2013.

——, 2015. 'NIP in Space Port Talks.' *Post-Courier*, 8 January 2015.

——, 2022a. 'Wesley Wants Govt to Stop Reported Sale of Island in Milne Bay.' *The National*, 30 August 2022.

——, 2022b. 'Gov't Keen on Protecting Island Group.' *The National*, 1 September 2022.

Aute, K., 2005. 'Clarify Fishing Deal.' *The National*, 9 May 2005.

Belshaw, C., 1955. *In Search of Wealth: A Study of the Emergence of Commercial Operations in the Melanesian Society of Southeastern Papua*. Memoirs of the American Anthropological Association (Memoir 80).

Berde, S., 1983. 'The Impact of Colonialism on the Economy of Paneati.' In J.W. Leach and E. Leach (eds), *The Kula: New Perspectives on Massim Exchange*. Cambridge: Cambridge University Press.

Blaikie, P. and H. Brookfield (eds), 1987. *Land Degradation and Society*. London: Routledge.

Bourke, R.M. and T. Betitis, 2003. 'Sustainability of Agriculture in Bougainville Province, Papua New Guinea.' Canberra: The Australian National University, Research School of Pacific and Asian Studies, Department of Human Geography.

Butler, J.R.A., T. Skewes, D. Mitchell and others, 2014. 'Stakeholder Perceptions of Ecosystem Service Declines in Milne Bay, Papua New Guinea: Is Human Population a More Critical Driver Than Climate Change?' *Marine Policy* 46: 1–13. doi.org/10.1016/j.marpol.2013.12.011

Butler, J.R.A., W. Rochester, T.D. Skewes and others, 2020. 'How Feasible Is the Scaling-Out of Livelihood and Food System Adaptation in Asia-Pacific Islands?' *Frontiers in Sustainable Food Systems* 4: 43. doi.org/10.3389/fsufs.2020.00043

Carrier, A.H. and J.G. Carrier, 1991. *Structure and Process in a Melanesian Society: Ponam's Progress in the Twentieth Century*. Chur: Harwood Academic Publishers.

Carrier, J.G., 1981. 'Labour Migration and Labour Export on Ponam Island.' *Oceania* 51: 237–255. doi.org/10.1002/j.1834-4461.1981.tb01469.x

Carrier, J.G. and A.H. Carrier, 1989. *Wage, Trade and Exchange in Melanesia*. Berkeley: University of California Press.

Christensen, A. and O. Mertz, 2010. 'Researching Pacific Island Livelihoods: Mobility, Natural Resource Management and Nissology.' *Asia Pacific Viewpoint* 51: 278–287. doi.org/10.1111/j.1467-8373.2010.01431.x

Cinner, J., M. Marnane, T. Clark and others, 2005. 'Trade, Tenure, and Tradition: Influence of Sociocultural Factors on Resource Use in Melanesia.' *Conservation Biology* 19: 1469–1477. doi.org/10.1111/j.1523-1739.2005.004307.x

Connell, J., 2010. 'Pacific Islands in the Global Economy: Paradoxes of Migration and Culture.' *Singapore Journal of Tropical Geography* 31: 115–129. doi.org/10.1111/j.1467-9493.2010.00387.x

——, 2016. 'Last Days in the Carteret Islands? Climate Change, Livelihoods and Migration on Coral Atolls.' *Asia Pacific Viewpoint* 57: 3–15. doi.org/10.1111/apv.12118

Connell, J. and N. Lutkehaus, 2017. 'Environmental Refugees? A Tale of Two Resettlement Projects in Coastal Papua New Guinea.' *Australian Geographer* 48: 79–95. doi.org/10.1080/00049182.2016.1267603

Dixon, R.D., 1981. *A Brief History of Mussau, Emira, and Tench Islands*. Morisset: Robert Dixon.

Egloff, B.J., 1978. 'The Kula before Malinowski: A Changing Configuration.' *Mankind* 11: 429–435. doi.org/10.1111/j.1835-9310.1978.tb00671.x

Epstein, A.L., 1969. *Matupit: Land, Politics, and Change among the Tolai of New Britain*. Canberra: Australian National University Press.

Eroro, S., 2008. 'Tides Hit 25,000.' *Post-Courier*, 11 December 2008.

Feinberg, R., 2009. 'Nukumanu Kinship and Contested Cultural Construction.' *Journal of the Polynesian Society* 118(3): 259–292.

Filer, C., 2014. 'The Double Movement of Immovable Property Rights in Papua New Guinea.' *Journal of Pacific History* 49: 76–94. doi.org/10.1080/00223344.2013.876158

——, 2017. 'The Formation of a Land Grab Policy Network in Papua New Guinea.' In S. McDonnell, M.G. Allen and C. Filer (eds), *Kastom, Property, and Ideology: Land Transformations in Melanesia*. Canberra: ANU Press. doi.org/10.22459/KPI.03.2017.06

——, 2019. 'Two Steps Forward, Two Steps Back: The Mobilisation of Customary Land in Papua New Guinea.' Canberra: The Australian National University, Crawford School of Public Policy, Development Policy Centre (Discussion Paper 86). doi.org/10.2139/ssrn.3502585

Filer, C., M. Andrew, B.Y. Imbun and others, 2016. 'Papua New Guinea: Jobs, Poverty, and Resources.' In G. Betcherman and M. Rama (eds), *Jobs for Development: Challenges and Solutions in Different Country Settings*. Oxford: Oxford University Press. doi.org/10.1093/acprof:oso/9780198754848.003.0004

Filer, C. and T. Wood, 2021. 'Geographical Constituents of Human Well-Being in Papua New Guinea: A District-Level Analysis.' Canberra: The Australian National University, Crawford School of Public Policy, Development Policy Centre (Discussion Paper 92). doi.org/10.2139/ssrn.3791993

Foale, S., 2005. 'Sharks, Sea Slugs and Skirmishes: Managing Marine and Agricultural Resources on Small, Overpopulated Islands in Milne Bay, PNG.' Canberra: The Australian National University, Research School of Pacific and Asian Studies, Resource Management in Asia-Pacific Program (Working Paper 64).

Fortune, R.F., 1963. *Sorcerers of Dobu*. London: Routledge and Kegan Paul.

Foster, R.J., 1995. *Social Reproduction and History in Melanesia: Mortuary Ritual, Gift Exchange, and Custom in the Tanga Islands*. Cambridge: Cambridge University Press.

Gabriel, J., P.N. Nelson, C. Filer and M. Wood, 2017. 'Oil Palm Development and Large-Scale Land Acquisitions in Papua New Guinea.' In S. McDonnell, M.G. Allen and C. Filer (eds), *Kastom, Property, and Ideology: Land Transformations in Melanesia*. Canberra: ANU Press. doi.org/10.22459/KPI.03.2017.07

Gharbaoui, D. and J. Blocher, 2017. 'Limits to Adapting to Climate Change through Relocations in Papua-New Guinea and Fiji.' In W.L. Filho and J. Nalau (eds), *Limits to Climate Change Adaptation*. New York: Springer. doi.org/10.1007/978-3-319-64599-5_20

Gosden, C. and C. Pavlides, 1994. 'Are Islands Insular? Landscape vs Seascape in the Case of the Arawe Islands, Papua New Guinea.' *Archaeology in Oceania* 29: 162–171. doi.org/10.1002/arco.1994.29.3.162

GPNG (Government of PNG), 2005. *National Gazette* #50, 13 April.

——, 2006. *National Gazette* #234, 28 December.

Guy, R., 1997. 'Traditional and Non-Traditional Mechanisms Designed to Alleviate Economic and Social Hardships on Wagifa Island, Milne Bay.' In R. Guy (ed.), *Formal and Informal Social Safety Nets in Papua New Guinea*. Port Moresby: National Research Institute.

Hair, C., S. Foale, J. Kinch and others, 2019. 'Socioeconomic Impacts of a Sea Cucumber Fishery in Papua New Guinea: Is There an Opportunity for Mariculture?' *Ocean and Coastal Management* 179: 104826. doi.org/10.1016/j.ocecoaman.2019.104826

Hanson, L.W., B.J. Allen, R.M. Bourke and T.J. McCarthy, 2001. *Papua New Guinea Rural Development Handbook*. Canberra: The Australian National University, Research School of Pacific and Asian Studies, Department of Human Geography.

Harding, T.G., 1967. 'Money, Kinship, and Change in a New Guinea Economy.' *Southwestern Journal of Anthropology* 23: 209–233. doi.org/10.1086/soutjanth.23.3.3629250

Hayes, G., 1993. '"MIRAB" Processes and Development on Small Pacific Islands: A Case Study from the Southern Massim, Papua New Guinea.' *Pacific Viewpoint* 34: 153–178. doi.org/10.1111/apv.342002

Hogbin, H.I., 1935. 'Trading Expeditions in Northern New Guinea.' *Oceania* 5: 375–407. doi.org/10.1002/j.1834-4461.1935.tb00162.x

——, 1970. *The Island of Menstruating Men: Religion in Wogeo, New Guinea*. Scranton: Chandler.

Irwin, G.J., 1985. *The Emergence of Mailu as a Central Place in Coastal Papuan Prehistory*. Canberra: The Australian National University, Research School of Pacific Studies, Department of Prehistory (Terra Australis 10).

Jackson, R.T., 1976a. 'Alotau and Samarai.' In R.T. Jackson (ed.), *An Introduction to the Urban Geography of Papua New Guinea*. Waigani: University of Papua New Guinea, Department of Geography (Occasional Paper 13).

——, 1976b. 'Daru.' In R.T. Jackson (ed.), *An Introduction to the Urban Geography of Papua New Guinea*. Waigani: University of Papua New Guinea, Department of Geography (Occasional Paper 13).

Kinch, J., 2020. Changing Lives and Livelihoods: Culture, Capitalism and Contestation over Marine Resources in Island Melanesia. Canberra: The Australian National University (PhD thesis).

Kurika, L.M., J.E. Moxon and M. Lolo, 2007. 'Agricultural Research and Development on Small Islands and Atolls: The Papua New Guinea Experience.' *Pacific Economic Bulletin* 22(3): 126–136.

Lauer, P.K., 1970. 'Amphlett Islands' Pottery Trade and the Kula.' *Mankind* 7: 165–176. doi.org/10.1111/j.1835-9310.1970.tb00403.x

Lipset, D., 2013. 'The New State of Nature: Rising Sea-levels, Climate Justice, and Community-Based Adaptation in Papua New Guinea (2003–2011).' *Conservation & Society* 11: 144–158. doi.org/10.4103/0972-4923.115726

Löffler, E. (ed.), 1977. *Geomorphology of Papua New Guinea.* Canberra: Commonwealth Scientific and Industrial Research Organisation and Australian National University Press.

Luetz, J. and P.H. Havea, 2018. '"We're Not Refugees, We'll Stay Here until We Die!": Climate Change Adaptation and Migration Experiences Gathered from the Tulun and Nissan Atolls of Bougainville, Papua New Guinea.' In W.L. Filho (ed.), *Climate Change Impacts and Adaptation Strategies for Coastal Communities.* New York: Springer. doi.org/10.1007/978-3-319-70703-7_1

Lutkehaus, N.C., 1985. 'Pigs, Politics and Pleasure: Manam Perspectives on Trade and Regional Integration.' *Research in Economic Anthropology* 7: 123–144.

———, 1995. *Zaria's Fire: Engendered Moments in Manam Ethnography.* Durham: Carolina Academic Press.

MA (Millennium Ecosystem Assessment), 2005. *Ecosystems and Human Well-Being: Synthesis.* Washington DC: Island Press.

Macintyre, M., 1983. 'Changing Paths: An Historical Ethnography of the Traders of Tubetube'. Canberra: The Australian National University (PhD thesis).

Macintyre, M. and J. Allen, 1990. 'Trading for Subsistence: The Case from the Southern Massim.' In D.E. Yen and J.M.J. Mummery (eds), *Pacific Production Systems: Approaches to Economic Prehistory.* Canberra: The Australian National University, Research School of Pacific Studies, Department of Prehistory (Occasional Paper 18).

Macintyre, M. and M. Young, 1982. 'The Persistence of Traditional Trade and Ceremonial Exchange in the Massim.' In R.J. May and H. Nelson (eds), *Melanesia: Beyond Diversity.* Canberra: The Australian National University, Research School of Pacific Studies.

Maina, J., J. Kithiia, J. Cinner and others, 2016. 'Integrating Social–Ecological Vulnerability Assessments with Climate Forecasts to Improve Local Climate Adaptation Planning for Coral Reef Fisheries in Papua New Guinea.' *Regional Environmental Change* 16: 881–891. doi.org/10.1007/s10113-015-0807-0

Martin, K., 2013. *The Death of the Big Men and the Rise of the Big Shots: Custom and Conflict in East New Britain.* New York: Berghahn Books.

McAlpine, J.R. and G. Keig, 1983. *Climate of Papua New Guinea.* Canberra: Commonwealth Scientific and Industrial Research Organisation and Australian National University Press.

Mirou, N., 2011a. Commission of Inquiry into SABL: Transcript of Proceedings at Kavieng, 26 October.

——, 2011b. Commission of Inquiry into SABL: Transcript of Proceedings at Kavieng, 28 October.

——, 2013. *Commission of Inquiry into Special Agriculture and Business Lease (C.O.I. SABL): Report.* Port Moresby: Government of Papua New Guinea.

Moyle, R., 2007. *Songs from the Second Float: A Musical Ethnography of Tāku Atoll, Papua New Guinea.* Honolulu: University of Hawai'i Press.

Mueller, A., 1972. 'Notes on the Tulun or Carteret Islands.' *Journal of the Papua New Guinea Society* 6(1): 77–83.

Munn, N.D., 1992. *The Fame of Gawa: A Symbolic Study of Value Transformation in a Massim Society.* Durham: Duke University Press.

Narayn, S., R.J. Cuthbert, E. Neale and others, 2015. 'Protecting against Coastal Hazards in Manus and New Ireland Provinces Papua New Guinea: An Assessment of Present and Future Options.' Goroka: Wildlife Conservation Society.

Nevermann, H., 1933. *St Matthias-Gruppe. Ergebnisse der Sud-See-Expedition 1908–1910.* Hamburg: Friederichsen, de Gruyter & Co.

Nicholas, I., 2008. 'Huge Waves Leave Many New Irelanders Homeless.' *The National,* 10 December 2008.

O'Collins, M., 1990. 'Carteret Islanders at the Atolls Resettlement Scheme: A Response to Land Loss and Population Growth.' In J. Pernetta and P. Hughes (eds), *Implications of Expected Climate Changes in the South Pacific Region.* Nairobi: United Nations Environment Programme.

Oram, N.D., 1976. *Colonial Town to Melanesian City: Port Moresby 1884–1974.* Canberra: Australian National University Press.

Paniu, L., 2011. 'LOs Were Not Given Explanation.' *Post-Courier*, 20 September 2011.

Philemon, T., 2022. 'Stop Conflict Islands Sale.' *The National*, 6 September 2022.

Philip, N.K., 2008. 'Project Emirau to Boost Growth.' *The National*, 9 September 2008.

Pomponio, A., 1992. *Seagulls Don't Fly into the Bush: Cultural Identity and Development in Melanesia*. Belmont: Wadsworth.

——, 1993. 'Education is Development on a Ten-Acre Island.' In V.S. Lockwood, T.G. Harding and B.J. Wallace (eds), *Contemporary Pacific Societies: Studies in Development and Change*. Englewood Cliffs: Prentice Hall.

Purdy, D.H., D.J. Hadley, J.O. Kenter and J. Kinch, 2017. 'Sea Cucumber Moratorium and Livelihood Diversity in Papua New Guinea.' *Coastal Management* 45: 161–177. doi.org/10.1080/08920753.2017.1278147

Rakova, U., 2014. 'The Sinking Carteret Islands: Leading Change in Climate Change Adaptation and Resilience in Bougainville, Papua New Guinea.' In S. Leckie (ed.), *Land Solutions for Climate Displacement*. London: Routledge.

Rasmussen, A.E., 2015. *In the Absence of the Gift: New Forms of Value and Personhood in a Papua New Guinea Community*. New York: Berghahn Books. doi.org/10.2307/j.ctt9qdb0f

Robbins, J., 2004. *Becoming Sinners: Christianity and Moral Torment in a Papua New Guinea Society*. Berkeley: University of California Press.

Schneider, K., 2012. *Saltwater Sociality: A Melanesian Island Ethnography*. New York: Berghahn Books.

Schwartz, T., 1963. 'Systems of Areal Integration: Some Considerations Based on the Admiralty Islands of Northern Melanesia.' *Anthropological Forum* 1: 56–97. doi.org/10.1080/00664677.1963.9967181

Siko, N., 2004. 'Cult-Like Storm Brewing on Island.' *Post-Courier*, 22 December.

Smith, M.F., 1994. *Hard Times on Kairiru Island: Poverty, Development, and Morality in a Papua New Guinea Village*. Honolulu: University of Hawai'i Press. doi.org/10.1515/9780824843267

Terrell, J.E. and R.L. Welsch, 1990. 'Trade Networks, Areal Integration, and Diversity along the North Coast of New Guinea.' *Asian Perspectives* 29: 156–165.

Thistleton, R., 2014. 'Want to Buy an Island?' *Sydney Morning Herald*, 14 January 2014.

Tiwari, S., 2012. 'Manus Island Sinking at Faster Rate than Carterets.' *The National*, 19 September 2012.

Vuvu, E., 2008. 'Islanders Experience Shortage.' *The National*, 11 April 2008.

West, P., 2016. *Dispossession and the Environment: Rhetoric and Inequality in Papua New Guinea*. New York: Columbia University Press. doi.org/10.7312/west17878

Whiting, N., 2021. 'Locals in Papua New Guinea Speak Out as China's Proposed Industrial Fishing Park Sets Off Alarm Bells in Canberra.' *ABC News*, 10 February 2021.

3

Livelihood Dilemmas on Some Small Islands in Milne Bay Province, Papua New Guinea

Simon Foale, Colin Filer, Jeff Kinch and Martha Macintyre

This chapter begins by exploring the way in which one group of small islands—the Bwanabwana group—became a focus of ethnographic attention during the brief period, from 2004 to 2006, when the Milne Bay Community-Based Coastal and Marine Conservation Project (MBCP) was being implemented by Conservation International. The rather limited nature of this attention and intervention is placed in the longer historical context of what is known about the lives and livelihoods of the islanders before they became the subjects of an externally funded conservation project that failed to achieve its own objectives. We then proceed to document what little we know about the further transformation of their lives and livelihoods in the wake of this failure. This leads us to reflect on some of the larger questions raised by the divergence of scientific and indigenous beliefs and practices relating to the conservation or exploitation of marine resources, both in Milne Bay Province and in other parts of Papua New Guinea (PNG).

The authors of this chapter have contributed to the findings in different ways. Martha Macintyre and Jeff Kinch have both undertaken long periods of fieldwork on particular islands in Milne Bay Province, while Simon Foale and Colin Filer have only been engaged, for brief periods, in the design and implementation of specific studies that were more or less connected with the MBCP and its aftermath. Macintyre's fieldwork was mostly conducted on one of the islands in the Bwanabwana group (Tubetube), while Kinch's fieldwork was conducted on another island (Brooker) that is not part of the Bwanabwana group but is not too far away. Kinch was involved in the design of the MBCP because of his prior experience as an anthropologist in the area of interest, but was not part of the team that attempted to implement the project. Kinch and Foale have both been involved in the design and evaluation of another project, in another part of PNG, which is discussed towards the end of this chapter because the outcomes bear some comparison with the outcomes of the MBCP. Given the divergence in the nature of our contributions, we have generally avoided the use of the first person plural in the remainder of the chapter except where it serves to signal our agreement on the point being made.

Frameworks, Scales and Zones

As indicated in Chapter 1, the Milne Bay Small Islands in Peril (SMIP) Program was originally conceived in 2001 as a component of the MBCP. The program was included in the final design documents for the MBCP that were submitted to the Global Environment Facility in January 2002 and approved in May of that year.

The SMIP Program was meant to contribute to the fourth output of the first phase of the MBCP, which was that '[p]olicies on sustainable development and land use strategies for densely populated small islands are finalised, reflecting the nexus between environment, poverty and governance' (GPNG and UNDP n.d.: 6). It was not entirely clear whose policies might be 'finalised' within a five-year time frame, but the specific objectives of the SMIP Program were itemised as follows:

- To build a credible and feasible framework for the collection, analysis and synthesis of ecosystem-wide data for decision-making at the level of the local community and the province as a whole.
- To test this framework in community-based assessments of ecosystem services in the area(s) of interest to the MBCP.
- To address decision-making information needs at the provincial level by means of scientific analysis, scenario construction and policy advice.
- To build capacity to undertake integrated assessments of the relationship between ecosystems and socio-economic systems at local, provincial and national scales.

ECONOMIC, SOCIAL AND CULTURAL FEATURES
OF ISLAND AND COASTAL COMMUNITIES

INNOVATION

MATERIAL IMPORTS
AND INFLOWS

CONSUMPTION
AND
MANAGEMENT
DECISIONS

EMIGRATION

MATERIAL EXPORTS
AND OUTFLOWS

GOODS AND SERVICES (OR BASIC NEEDS) SUPPLIED
BY TERRESTRIAL AND MARINE ECOSYSTEMS

EXTERNAL DRIVERS AND CONSTRAINTS

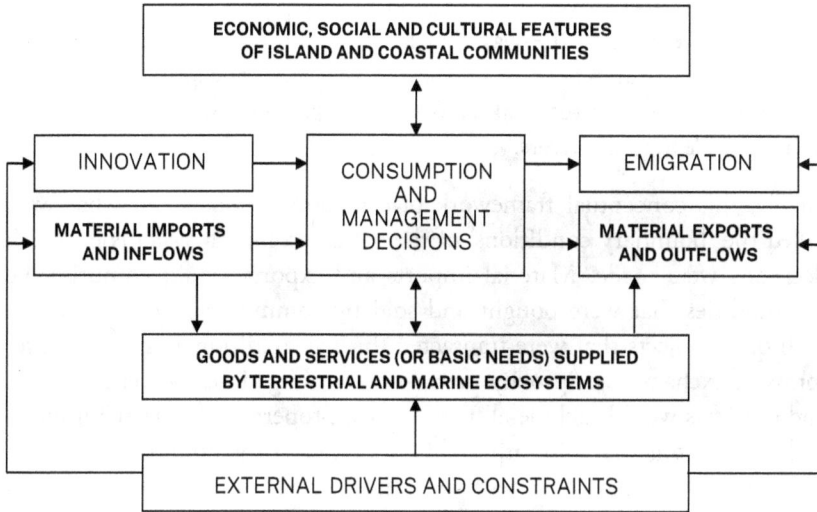

Figure 3.1: Initial conceptual framework for the Milne Bay SMIP Program
Source: Filer 2002: 6.

The SMIP Program was understood from the outset to be a local or community-based 'ecosystem assessment' that would be part of the Millennium Ecosystem Assessment (MA). In 2001, it was already clear that it would need a conceptual framework that was broadly consistent with the one to be adopted by this global assessment exercise, but this exercise was still a work in progress. So Colin Filer came up with one that appeared to match what was already known about the circumstances of 'densely populated small islands' in Milne Bay Province, but which could also be seen as 'a model of the interaction between ecosystems and socio-economic systems' at different geographical scales or levels of political organisation (Figure 3.1).

This conceptual framework diverged in a number of ways from the one that was formally adopted by the MA in 2003 (MA 2003: 37). Instead of placing the relationship between ecosystem services and human well-being at the centre of the picture, it reserved this central location for the 'consumption and management decisions' taken by members of local communities, or possibly by other actors at higher levels of political organisation. These were taken to include decisions about human reproduction, and hence the rate of population growth, as well as decisions about the consumption and management of natural resources (or ecosystem goods and services). This conceptual framework paid less attention to the distinction between direct and indirect drivers of change in the supply of ecosystem services, and more attention to the way that local consumption and management decisions, which are called 'responses' in the MA conceptual framework, are influenced by institutional factors that might be described as 'internal' rather than 'external' drivers.

Finally, this conceptual framework placed more emphasis on what were called the 'boundary conditions' of the social–ecological systems in which decisions were made. Material imports and exports would comprise the commodities that were bought and sold by community members, along with other objects that were transacted through traditional or non-market forms of exchange, reciprocity or customary obligations. Material inflows and outflows would include all the physical properties of material imports and exports, together with other 'things' that cross the boundaries of small island ecosystems, with or without the knowledge or encouragement of community members, like the African snails (*Achatina fulica*) that were invading and devouring local gardens, or the greenhouse gas emissions from the operation of outboard motors. The idea of a balance between 'emigration' and 'innovation' was based on the familiar choice faced by small island communities in which population growth places increasing pressure on available resources or ecosystem services. While the rate of emigration was likely to exceed the rate of immigration in these communities, it should not be assumed that this was invariably the case.

A conceptual framework does not constitute a hypothesis, and this one was simply intended as the first step towards the achievement of the SMIP Program's first objective, since it would guide the 'collection, analysis and synthesis of ecosystem-wide data for decision-making at the level of the local community' and the province as a whole (Filer 2002: 5). But where would the framework be applied to the actual collection, analysis and synthesis of the data?

It was already known that Milne Bay Province contained about one-third of all the small islands in PNG that were initially identified as being under pressure or in peril because of their crude population densities at the turn of the millennium. It was also known to contain a more extensive set of coral reef ecosystems than any other coastal or maritime province in PNG, and was, therefore, a natural magnet for conservationists looking to escape from the shadows of their failure to convince PNG's 'rainforest people' to resist the temptations of large-scale resource development. However, these small islands and coral reef ecosystems were widely distributed within the province, so geographical priorities would have to be established.

The design of the MBCP divided the province into four zones and set out a plan to mobilise provincial and local stakeholders to support the establishment of a network of community-managed marine protected areas in Zone One during the first (five-year) phase of the program, and then to extend this effort to Zones Two and Three during the second (three-year) phase. In the period that elapsed between the initial design of the MBCP and the abbreviated implementation of the first phase between 2004 and 2006, there was some debate about whether the boundaries of these four zones should be demarcated by reference to the distribution of coral reef ecosystems or should match the political and administrative boundaries of districts and local-level government (LLG) areas (Allen et al. 2003; van Helden 2004; Baines et al. 2006). By 2004, Zone One encompassed the whole of Alotau District and two of the four LLG areas in Samarai–Murua District, which meant that it comprised approximately one third of the population of the entire province. However, many of the communities in Alotau District are not coastal communities, and many of the coastal communities are not associated with a coral reef ecosystem ostensibly containing the sort of biodiversity values that would merit the establishment of a protected area (Foale et al. 2016).

Some attempt had been made to develop criteria by which specific coastal and small island communities would be selected for the process of 'community engagement' (Kinch 2001), but the eventual choice appears to have been made by individual program staff for reasons of their own (Baines et al. 2006). As implementation of the first phase of the MBCP was compressed into a period of less than three years, the staff concentrated most of their efforts on a group of just three small island communities in the Maramatana LLG area, which is part of Alotau District. Since the few hundred members of these communities were somewhat overwhelmed by the sudden burst of attention from what amounted to a whole community of conservationists based in the provincial capital, the SMIP Program was pointed in a different direction in order to avoid any addition to this new source of pressure. And that is how the SMIP Program's conceptual framework came to be applied to the collection, analysis and synthesis of information about the way that resource management decision were being made in a different collection of small island communities within the boundaries of Zone One.

The Bwanabwana Language-Island Group

The name Bwanabwana is applied to a language, to the people who speak that language, to the islands that they inhabit and to the LLG in which the islanders are represented.[1] This can be a source of some confusion, since the Bwanabwana people or speakers account for less than 25 per cent of the people represented in the LLG of the same name.[2] Here we use the name in its narrower sense, to refer to the people who speak the language or the islands that they occupy. The islands that they occupy are all very small or miniscule, according to the criteria adopted in Chapter 2, since none has a surface area of more than 10 km^2, and some have a surface area of less than 1 km^2. They are part of the Louisiade Archipelago and lie to the east of the medium-sized island of Basilaki (or Bwasilaki), which is itself part of the LLG area but is occupied by people who speak other languages. Most of the Bwanabwana islands were collectively known as the 'Engineer Group' during the colonial period because they were given the names of the engineers or crew members of HMS *Basilisk*, under the command of Captain John Moresby, when it sailed through the area in 1873. Some of these 'engineering' names are still present in the national census or other government records, while others have largely been forgotten. The Bwanabwana people have their own names for each of the islands in the group, and these are the names that will generally be preferred in the following discussion (see Figure 3.2).

At the centre of this cluster of islands is a string of three very small (but not miniscule) islands whose local names are Tubetube, Naluwaluwali and Kwaraiwa, and whose engineering names were Slade, Skelton and Watts. Tubetube and Kwaraiwa (sometimes spelt Kwalaiwa) are generally known by their local names, but Naluwaluwali (which literally means 'Their place in the middle') is still called 'Skeleton Island' in the national census, following the apparent corruption of its original engineering name, Skelton.[3] Tubetube and Naluwaluwali belong to a single council ward that also bears the name of Tubetube, while Kwaraiwa is a separate council ward.

---

1   In documents relating to the design of the MBCP, the name was also applied to the whole of Zone One, including the whole of Alotau District, which made no political sense to any of the local stakeholders.

2   The national census counted 8,894 citizens as residents of the Bwanabwana LLG area in 2000.

3   Naluwaluwali is also shown on some maps with the alternative local name of Naunalualua, which should probably be spelt Nuanaluwaluwa to accord with local usage. The term *luwaluwa* ('in the middle') occurs in both versions of the local name.

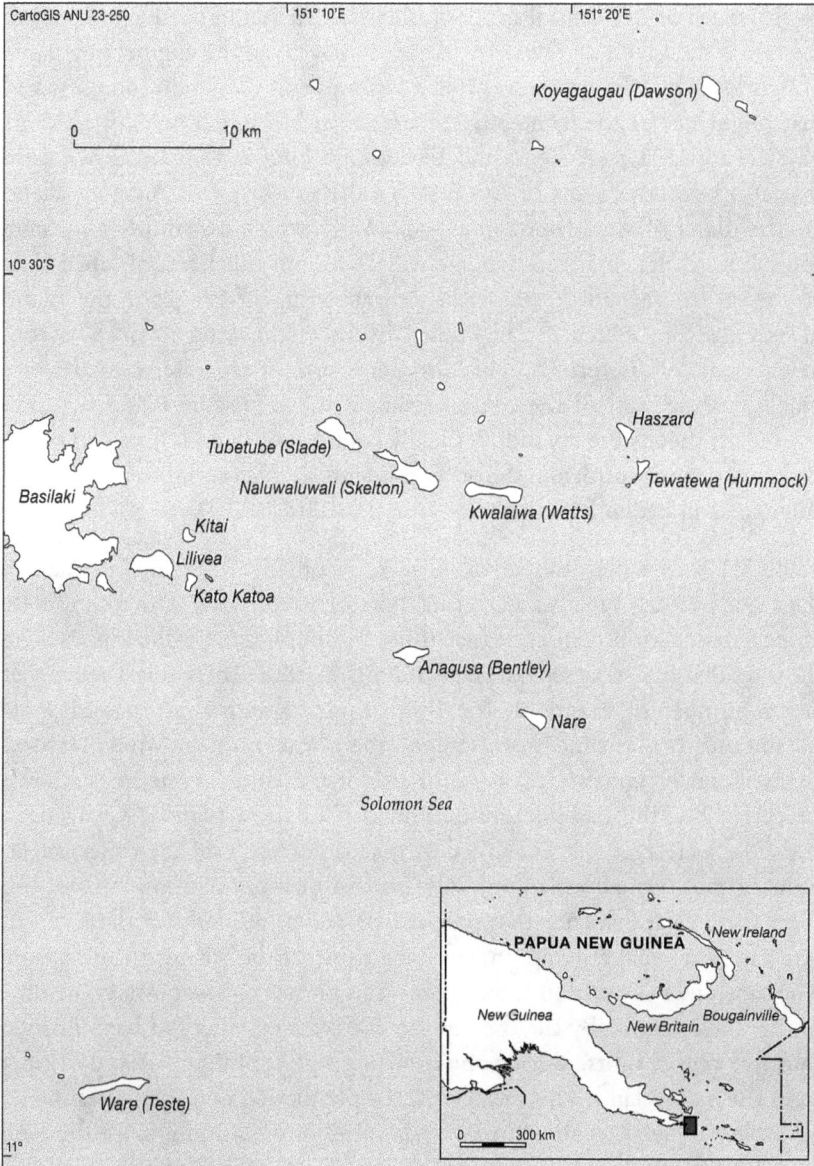

Figure 3.2: Map of the Bwanabwana region
Source: CartoGIS Services, College of Asia and the Pacific, The Australian National University.

To the south of this central string of islands is the island of Anagusa, and to the east is the island of Tewatewa. Anagusa was given the engineering name of Bentley, while Tewatewa was called Hummock by Captain Moresby, and that might not be the name of a crew member but rather an allusion to its physical appearance. Anagusa and Tewatewa belong to a single council ward that also bears the name of Anagusa. To the southwest of Anagusa Island lies the island of Ware (sometimes spelt Wari), which constitutes a separate council ward. Its engineering name was Teste, but this has long since been forgotten. To the northeast of the central string of islands is the island of Koyagaugau, which is still known by its engineering name, Dawson, in national census records. This also constitutes a separate council ward. Finally, to the west of the central string is the miniscule island of Kitai, located off the eastern coast of Basilaki Island. Kitai Island is now part of the Bedauna council ward, and the other residents of this ward speak the Tawala language and live in Bedauna village on Basilaki Island.[4]

Table 3.1 shows the number of villagers recorded as residents of each of the eight Bwanabwana island census units in 2000, along with an estimate of their population density at that time. Although these eight islands were the ones designated as separate census units in 2000, there was a somewhat larger number of islands in the Bwanabwana group that boasted some inhabitants at that time. For example, the census unit known as Dawson Island actually consists of a group of three islands, sometimes known collectively as the Laseinie group, and two of these islands, Koyagaugau and Ole, were certainly inhabited at that time. Some of the other census units in the Bwanabwana group also consist of more than one island, but there is no clear evidence that people recorded as residents of these census units were normally living on more than one of the islands in the group. The most likely exception is the census unit known as Kwaraiwa at the time of the 2000 census. By the time of the 2011 census, this had been divided into five census units, at least one of which was a miniscule island distinct from the main island. This means that the population density figures shown in Table 3.1 need to be treated with a measure of caution, since the land areas specified in that table refer to the size of islands that were known to be inhabited in 2000.

---

4   It is not clear whether this is still a bilingual council ward or whether one of the two languages has now come to predominate.

Table 3.1: Bwanabwana islands recorded in the 2000 national census

Island census unit	Residents in 2000	Land area (km²)	Density in 2000
Koyagaugau (Dawson)	187	1.4	133.6
Tubetube	184	2.4	76.7
Naluwaluwali (Skeleton)	282	2.8	100.7
Kwaraiwa	358	1.9	188.4
Tewatewa	78	0.6	130.0
Anagusa	85	1.3	65.4
Ware	707	2.2	321.4
Kitai	177	0.5	354.0
TOTAL	2,058	13.1	157.1

Source: PNG national census data.

Table 3.2 shows the growth of the resident population of each of the eight island census units over the period in which they have been counted in the national census. Although the most recent (2011) census is said to have been a failure at the national level, the figures for small island populations in Milne Bay Province are probably as accurate as those collected in previous headcounts. Table 3.2 shows some very high rates of population growth between 2000 and 2011 on some of the islands in the group—most notably Tubetube and Ware. As we shall see, this is most likely explained by the boom in the bêche-de-mer fishery that took place during the first half of this period.

Table 3.2: Growth of the Bwanabwana island population, 1980–2011

Island census unit	1980	1990	2000	2011
Koyagaugau & Ole	118	141	187	247
Tubetube	104	224	184	311
Naluwaluwali	132	138	282	256
Kwaraiwa	218	296	358	474
Tewatewa	59	58	78	91
Anagusa	70	72	85	119
Ware	498	558	707	1,250
Kitai	99	142	177	183
TOTAL	1,298	1,629	2,058	2,931

Source: PNG national census data.

Two of the baseline studies conducted for the MBCP included some information on the lives and livelihoods of the people resident on this group of islands at the time of the 2000 national census (Kinch 2001; Mitchell et al. 2001). Given the prospective focus of the MBCP on the establishment of marine protected areas, much of this information related to the exploitation of marine resources by the islanders, but some of it related to other economic activities or access to public goods and services. The first social scientist to gather such information was anthropologist Cyril Belshaw, who conducted fieldwork on Ware Island in 1951 (Belshaw 1955). Forty years later, demographer Geoff Hayes conducted three successive household surveys on the same island (Hayes 1993). Meanwhile, anthropologist Martha Macintyre had conducted several months of fieldwork on Tubetube Island between 1979 and 1981, and subsequently made a number of return visits to the same island (Macintyre 1983, 1987, 1989). She also travelled to other islands in the Bwanabwana group as part of her investigation of social and economic relationships between the different island communities (Macintyre and Young 1982; Macintyre and Allen 1990). A survey of local agricultural systems was undertaken in 1994 as part of the nationwide Mapping Agricultural Systems Project funded by the Australian aid program (Hide et al. 2002).

From these sources, it is clear that periodic droughts are a major threat to local livelihoods. Macintyre (1983) observed that crops failed on average once every decade. She witnessed the effects of a severe drought in 1981, and this was followed by events of equal severity in 1991, 1997 and 2015 (Kinch 2020).[5] As noted in Chapter 2, Milne Bay Province is also vulnerable to the occasional cyclone, though not to the same extent as some Pacific island nations (McAlpine and Keig 1983; Skewes et al. 2011). The maritime trading networks that traditionally connected the Bwanabwana islands with other coastal communities helped to mitigate the locally uneven effects of such extreme weather events (Macintyre and Allen 1990). That is not only because they enabled some islands to specialise in the production of artefacts that could be exchanged for foodstuffs, but also because they enabled the people themselves to circulate between communities at various intervals (Macintyre 1983). Elements of this traditional form of circulation are still present, but have now been supplemented by more distinctively modern forms of migration in what Hayes (1993: 166) described as a 'two-circuit system of mobility'.

---

5   Milne Bay Province also had unusually long dry seasons in 2005, 2009 and 2013.

This might seem to be at odds with the observation made in Chapter 2 about the unusually low proportion of Milne Bay island populations recorded as being absent from their island homes at the time of the 1980 census. However, the absentee rates were considerably higher in the Bwanabwana islands than in other parts of the province (see Table 3.3). In his own demographic survey, Hayes (1993) found that almost half of the household heads on Ware Island had spent part of their lives living somewhere else, either in another rural community or in an urban centre. Hayes was primarily interested in the extent to which the modern form of migration helped to sustain the livelihoods of people still living on Ware Island through the delivery of remittances by community members living elsewhere. He found that remittances in cash accounted for roughly one-third of the average household income of 660 kina per annum in the early 1990s,[6] but was unable to measure the value of additional remittances in kind or 'cargo' that migrants brought back to the island. He also reckoned that subsistence production based on the harvest of terrestrial and marine resources supplied an average of about one-third of the food required by the islanders living at home, although the composition of this package of ecosystem services would have varied from year to year because of changes in the weather.

Table 3.3: Absentees in the Bwanabwana island population, 1979–1980

Island census unit	Residents in 1980	Absentees in 1979	Absentee rate (%)
Koyagaugau (Dawson)	118	10	7.8
Tubetube	104	26	20.0
Naluwaluwali (Skeleton)	132	26	16.5
Kwaraiwa	218	34	13.5
Tewatewa	59	17	22.4
Anagusa	70	26	27.1
Ware	498	95	16.0
Kitai	99	22	18.2
TOTAL	1,298	256	16.5

Source: PNG provincial data system and national census data.

---

6   At that time one PNG kina was worth about the same as one US dollar.

Aside from remittances, the islanders were able to make up the balance through the export of various goods and services. Changes in the composition of this package of exports had much less to do with climatic cycles than with medium- and long-term changes in the market for different commodities. In the early 1950s, the Ware Islanders were making money by cutting copra, building boats and selling or trading the pots that were an island speciality (Belshaw 1955). By the early 1990s, the export of trochus shells accounted for 15 per cent of average household incomes, while the export of bêche-de-mer, the sale of local pottery, and boat charters (rather than boat construction) each accounted for roughly half that proportion (Hayes 1993). By 1991, only one household was still selling copra, and production of this commodity had ceased altogether by 1994 (Hide et al. 2002). Copra was still the most significant of the commodities exported from the Bwanabwana islands in the early 1980s, but trochus and bêche-de-mer were more significant sources of cash income throughout the region by the mid-1990s, when surplus coconuts from senile plantations were more likely to be fed to the local pig population (Macintyre 1983; Hide et al. 2002). Copra production from islands aside from Ware has been sporadic since the 1990s as the price of copra has fluctuated from one year to the next (Foale 2005; Kinch 2020).

However, there is more to the economic and social history of these islands than a tale of the rise and fall of different commodities, even if the definition of a commodity is expanded to include the wide variety of products that passed through traditional trading networks, as well as those bound for the world market (Macintyre 1983; Macintyre and Allen 1990). Martha Macintyre's investigation of the historical record indicates that the population of Tubetube Island when Methodist missionaries first arrived in 1892 was almost four times the population that was living there at the time of the 1980 census, while the populations of the neighbouring islands of Naluwaluwali and Kwaraiwa were less than half the size of the populations counted in 1980. She ascribed this demographic divergence to an erosion of the relative power of Tubetube, both as a trading hub and a community of warriors, following the imposition of colonial rule (Macintyre 1983: 21–2). If that is the case, it would exemplify the need to treat the social and cultural features of island communities, and the distribution of power and influence within a network of such communities, as an independent variable in accounting for the way that islanders manage the consumption of ecosystem services or for patterns in the circulation of material objects and their human producers (see Figure 3.1). The decisions made by island households therefore reflect a mixture of environmental, economic and institutional factors that are not easily disentangled.

The Bêche-de-Mer Boom

In the baseline studies undertaken for the MBCP it was already clear that the main challenge to the success of the project would stem from the reliance of local communities on the cash incomes obtained from the rapid depletion of specific marine resources or 'ecosystem services'. The most significant of these were the various species of sea cucumber, known in Milne Bay as *buvoki*, that were being extracted from shallow coastal waters, then processed onshore to become the commodity known as bêche-de-mer.[7] The commodity was then purchased by local trading companies that exported most of it to buyers based in Hong Kong, with smaller volumes traded to Singapore and Malaysia. Chinese traders had been visiting the Bwanabwana islands to purchase bêche-de-mer back in the nineteenth century (Russell 1970), but the local trade in this commodity was quite limited for most of the century that followed because it has no value beyond the limits of the Asian markets. The recent boom began in the 1990s, as a rapid growth in demand from China led to an equally rapid increase in the prices paid to local producers and traders. By 1997, bêche-de-mer accounted for 28 per cent of the value of all marine commodity exports from Milne Bay Province, whereas trochus accounted for 25 per cent (Mitchell et al. 2001). However, the growing predominance of bêche-de-mer was already much greater in some parts of the Louisiade Archipelago, including the Bwanabwana islands (Kinch 2001).

By 2001, Milne Bay Province alone accounted for 43 per cent of the total value of PNG's exports, which meant that the province obtained an income of 7.8 million kina from the export of 210 metric tonnes of bêche-de-mer (Kinch 2002). By that time, prices were 30 times higher than they had been in 1990 (Kinch et al. 2008).[8] But the windfall had come at an ecological price. A stock assessment in 2001 found that sea cucumber populations had

---

7   Most sea cucumbers belong to the class Holothuroidea in the phylum Echinodermata. Strictly speaking, bêche-de-mer is the dried body wall of the sea cucumber, but the former name is sometimes applied to the living organism.

8   This was partly because the value of the kina had fallen substantially in comparison to the value of the US dollar.

fallen by 50 per cent over the previous decade, that population densities were already well below those found in the Torres Strait, and the extent of the decline was proportional to the market value of different species (Skewes et al. 2002). At that juncture, the two most valuable species— sandfish (*Holothuria scabra*) and black teat (*H. whitmaei*)—had already been harvested to the point of virtual extinction. Fishers had moved on to lower value species that were still relatively abundant. The boom of the 1990s therefore seemed to entail what Kinch (2002) described as the transformation of a low-volume high-value fishery into a high-volume low-value fishery, or what others have described as 'fishing down the price list' (Scales et al. 2006). However, the prices paid for low-value species towards the end of the boom could still match or exceed the prices paid for high-value species when the boom started.

The communities engaged in this form of commodity production responded to the growing scarcity of the resource in two ways: they intensified their fishing effort by adopting new technologies while seeking to establish more exclusive property rights over the ecosystems that harboured the resource (Foale et al. 2011). The process of technical innovation began with the substitution of fibreglass dinghies and outboard motors for traditional sailing canoes (known locally as sailaus),[9] and continued with the deployment of underwater diving gear to reach sea cucumbers at greater depths and electric torches to find them at night (Sabetian and Foale 2006). The owners of the dinghies were able to manipulate traditional ties of kinship and marriage between different communities to travel greater distances in search of fishing grounds that had not yet been depleted, but this, in turn, had the effect of creating new incentives for community leaders to demarcate and defend their maritime territorial boundaries against uninvited incursions, so the net result was an increase in the frequency and intensity of territorial disputes (Fabinyi et al. 2015; Kinch 2020).

By 2001 the National Fisheries Authority had introduced a national bêche-de-mer management plan in an effort to limit the damage that was being done. This established an annual 'total allowable catch' for each of the maritime provinces. The limit for Milne Bay Province was set at 140 tonnes, but the actual volume of exports in 2001 exceeded this limit by 50 per cent. The plan prohibited any harvest of sea cucumbers during the spawning season, between October and December, and prohibited

---

9   Sailaus are wooden-planked, single outrigger sailing canoes between four and 12 metres long (Smaalders and Kinch 2003).

the use of underwater diving gear and electric torches at any time of year, but both of these rules were flouted by some of the fishers in the province (Kinch 2002). The rules were still being flouted when implementation of the MBCP began in 2004.

Bwanabwana Livelihoods in 2005

Our own assessment of the relationship between ecosystem services and local livelihoods in the Bwanabwana islands involved a number of field trips between April 2005 and February 2006. The first field trip involved an assessment of local fishing practices and attitudes to marine resource management in six of the island communities—Tubetube, Naluwaluwali, Kwaraiwa, Ware, Anagusa and Koyagaugau. The second was intended to collect information on the incidence of disputes over access to marine ecosystems and resources. The third aimed to discover the way that the bêche-de-mer boom was affecting the traditional circulation of people and goods between island communities, with particular attention paid to the islands of Tubetube, Naluwaluwali, Ware and Koyagaugau. The final field trip involved a more detailed study of the harvesting of sea cucumbers by residents of Ware Island. Further inquiries should have continued until 2007, but the process of investigation was cut short by the evaporation of the MBCP budget, including the part that was earmarked for the SMIP Program (Baines et al. 2006; Balboa 2013).

We did not conduct the kind of household survey that Hayes had conducted on Ware Island in 1991 and 1992, so we did not manage to establish the distribution of cash incomes from different sources on each of the islands included in our sample. Indeed, we found that people were reluctant to share information about the income and expenditure of their own households, which may have reflected a more general concern about the incidence of economic inequality. However, it was quite clear that bêche-de-mer still accounted for most of the money that flowed through the local economy (Figure 3.3), although this was now supplemented by the incomes that some of the islanders were obtaining from the sale of dried shark fins. Evidence for this observation was obtained from interviews with the owners and operators of local trade stores, fibreglass dinghies and traditional sailing canoes.

Figure 3.3: Assortment of bêche-de-mer on Koyagaugau Island, 2005
Source: Photograph by Simon Foale.

Figure 3.4: Decoration of clay pot on Ware Island, 2005
Source: Photograph by Simon Foale.

By 2005 the open season for the sea cucumber fishery had been reduced to six months, from January to June, and closure of the fishery for the remainder of the year was having a considerable impact on a range of other economic activities—not just the volume of commodities imported and sold through local trade stores, but also the volume of goods exchanged through traditional trade between island communities. That was partly because the open season coincided with the time of year in which subsistence garden produce was in short supply, so more cash was required to purchase imported foodstuffs, while the period of closure left people with more time to engage in traditional forms of exchange, or for the Ware Islanders to compensate for the shortage of cash by selling their pots instead of selling bêche-de-mer (Figure 3.4).

There were 26 fibreglass dinghies distributed between the six islands that were surveyed in April 2005, though some appeared to be idle because they lacked an outboard motor. At that time, the average price of a new dinghy was 6,000 kina, while the average price of a new 40-horsepower motor was 9,000 kina.[10] To afford these prices, dinghy owners would commonly incur debts to the trading companies which then had to be paid off with earnings from the sale of bêche-de-mer and other marine commodities. Many of the dinghy operators travelled all the way to Alotau, the provincial capital, to sell their catch, returning with fuel, food and other goods that they purchased with the proceeds from their sales. In this respect, the dinghies had taken on some of the functions of the diesel-powered workboats that had long been the primary means of transporting people and goods between coastal towns and villages. However, the fuel efficiency of the dinghies was much lower than that of the workboats, which may help to explain why dinghies based on Ware Island were mainly used for fishing and their catch was then loaded onto workboats, a number of which were based on the island.

Ware Island is not just remote from other islands in the Louisiade Archipelago; it also has an unusually high population density and had an unusually high rate of population growth after the beginning of the bêche-de-mer boom. The extent of population pressure on the island's cultivable land was reflected in an apparent shortening of the fallow period in the local agricultural system from more than five years in 1994 to less than three in 2005 (Hide et al. 2002: 53; Foale 2005: 9). It might have been expected that this apparent degradation of the island's terrestrial resources would

---

10   At that time, one kina was worth about 45 Australian cents.

result in an intensification of land disputes between resident members of the island community. But what seems to have happened instead—or at least initially—is that the islanders simply intensified their exploitation of the marine resources available to them and invested more of the income from this harvest in the provision of public goods and services as well as the purchase of imported commodities to meet their household food requirements. Foale (2005) thought this association of increasing population pressure with a greater amount of 'social capital' marked a somewhat paradoxical difference between the Ware community and the other Bwanabwana island communities, but acknowledged the need for further research to establish the relationship between the ecological, economic and institutional dimensions of this difference.

The ecological dimension of the difference did not extend to a greater appreciation of the need for sustainable management of the marine resources on which the Ware Islanders were increasingly reliant, nor did the institutional dimension extend to a greater willingness to collaborate with members of other island communities in the establishment of a more inclusive management regime. From interviews conducted in the six island communities in April 2005:

> What emerged repeatedly was an overriding concern about the *rights of access* to resources, and relatively little concern about *sustaining* the income. Many people expressed annoyance at the use of [underwater diving] gear by a minority of fishers, reasoning that the increased access that this equipment gave to these men was unfair as it allowed them to make large amounts of money and at the same time remove [sea cucumbers] that would otherwise be fishable by the majority who were forced to dive on breath-hold. It was clear that people were mostly *not* thinking about a) the rate at which depleted resources would recover, b) how the process of stock recovery actually worked (i.e. spawning, fertilisation, larval dispersal and settlement), and c) the impact of removal of a given species or group of species on the rest of the ecosystem.
>
> (Foale 2005: 19, italics in original)

It could certainly be argued that the Ware community would secure a greater benefit from the establishment of a marine protected area, given the spatial extent of the marine ecosystems to which its members had access and the degree of their reliance on the sale of marine commodities to maintain their livelihoods. Furthermore, members of this community were more united in their opposition to the use of underwater diving gear to harvest sea cucumbers. However, this opposition could be explained by their desire to limit the opportunities for some individuals to profit at the expense of others who could not afford the new technology, or else by their desire to exclude members of other island communities from what they regarded as their own territorial waters.[11] There was little evidence to indicate that they or any of the other Bwanabwana islanders were prepared to make the consumption and management decisions that would halt the degradation of marine resources.

Aside from the incidence of extreme weather events, the main source of pressure on local livelihoods in the Bwanabwana islands was the increase in the resident population combined with a growing scarcity of starchy foods and raw materials for house construction. Additional starch was traditionally obtained by means of barter with neighbouring islands, but was now more likely to be purchased with the cash obtained by harvesting sea cucumbers and shark fins. The overharvesting of these marine resources had become another source of pressure, but the high prices paid for these marine commodities were partly responsible for the rate of population growth. Instead of responding to these pressures by imposing or accepting limits on the size of their catch, the islanders were taking out a form of insurance against the likelihood of future stock depletion by maintaining traditional social relationships with people on larger or less densely populated islands to keep open the option of resettlement. Our survey of four Bwanabwana islands in November 2005 found that nearly all the islanders thought they had an option to migrate to another island or coastal community in the event of a drought or severe depletion of currently lucrative marine resources. But this in turn meant that islanders had less incentive to take management decisions that would halt the process of depletion.

The future effect of this behaviour on the biodiversity values of coral reefs or other marine ecosystems was hard to determine. It could well be argued that the activities of foreign fishing vessels were a bigger threat to these values (Kinch 2001). However, the MBCP was driven to treat local people's attitudes and behaviour as the main source of pressure because it was a 'community-based' program, and therefore had no mandate to regulate the large-scale commercial fishing industry (van Helden 2004). At the same

---

11   Oddly enough, it was the Ware ward councillor who encouraged divers from Tubetube and Kwaraiwa islands to use underwater diving gear on a reef system that the Ware Islanders regarded as their own territory. It appears that the councillor had lost his own authority within his own community and had gone to live in the provincial capital.

time, the MBCP was locked into a managerial conceptual framework that made it very difficult to deal with a mixture of territorial and jurisdictional disputes, not only between neighbouring island communities or their individual members, but also between different levels of government and different agents or agencies at each level of political administration (Fabinyi et al. 2015; Kinch 2020). The conceptual framework adopted by the Millennium Ecosystem Assessment did not cast much light on this political dynamic (Filer 2009). However, it is not obvious that another conceptual framework would have made it any easier for an organisation like Conservation International to rein in the local entrepreneurs who were organising the export of the marine commodities. The former governor of Milne Bay Province, who had invested some of his own political capital in the MBCP, was defeated by one of these entrepreneurs in the 2007 national elections, and he had some reason to blame his defeat on the cancellation of the project that he had supported. But that was not the end of the political story, because the National Fisheries Authority imposed a nationwide moratorium on the harvest of sea cucumbers and the export of bêche-de-mer in October 2009.

School Fees, Fishing and Remittances

In March 2010, we investigated the effect of the bêche-de-mer moratorium on three of the Bwanabwana islands—Tubetube, Naluwaluwali and Ware. This involved a survey of 84 households spread across all three islands, semi-structured and unstructured interviews with many of the people included in the survey, and key informant interviews conducted with several bêche-de-mer fishers, two school headmasters (from Ware and Tubetube), two bêche-de-mer buyers and the Ware Island village enumerator.

Table 3.4 shows the primary source of income before and after the imposition of the moratorium. Prior to the moratorium, bêche-de-mer had been the primary source of income for 80 per cent of the households that we surveyed. Only 18 per cent of households had a second or third source of income while the fishery was open. After the moratorium, primary sources of income became more diversified, with copra being the most important (29 per cent of households), followed by clay pot manufacture (21 per cent, mostly on Ware), trochus (14 per cent) and shark fin (12 per cent). The proportion of households with a second or third source of income increased to 51 per cent, while the number with no source of income increased from one to six. The economic dominance of bêche-de-mer in the decade prior to our fieldwork was due to significant increases in prices for most species during that time (Kinch 2007), and contrasts with the more modest position it occupied in the Ware Island economy in 1991 and 1992, when only one third of households engaged in the fishery (Hayes 1993).

Table 3.4: Primary source of income before and after the moratorium

Income source	Before	After
Bêche-de-mer	67	0
Trochus	3	12
Fishing	2	3
Shark fin	1	10
Smoked fish	0	1
Copra	1	24
Pots	1	18
Tobacco, betelnut	1	0
Vegetable sales	0	2
Trade store	0	1
Family support	1	0
Government employment	1	1
Other employment	0	1
Remittances	3	3
Tithes	1	1
Superannuation	1	1
None	1	6
TOTAL	84	84

Source: Authors' interview data.

A key objective of our survey was to find out what proportion of the population believed the fishery was not in fact overharvested and would continue to provide substantial economic returns. Of the 64 survey respondents who answered the question, 'Do you agree with the moratorium?' 40 (62.5 per cent) said they disagreed, while 24 (37.5 per cent) said they agreed. The Ware village enumerator said that a lot more people agreed with the moratorium when it was first announced,[12] and the main reason for the change in their views was that they had since run out of money.

---

12 Staff of the National Fisheries Authority carried out an awareness campaign in affected areas when the moratorium was first announced.

Table 3.5: Reasons why people disagreed or agreed with the moratorium

Disagree	27
Income made people more equal; now there is more inequality	1
Main source of income	12
Can't pay school fees	10
Can't afford school fees, food, clothes	2
They should open it for abundant species	1
There are still plenty of bêche-de-mer (BDM)	1
Agree	22
To allow recovery	14
Awareness by National Fisheries Authority	1
Compressor users finished all the BDM — those of us in canoes had less to harvest	1
Resource is overharvested	5
Women doing less work in gardens because they could buy food with BDM money	1
TOTAL	49

Source: Authors' interview data.

Of the 40 who now disagreed with the moratorium, 27 gave a reason, and of these, just under half (12) cited the difficulty of paying school fees. One of the people who agreed with the moratorium also said that it was now harder to pay school fees (see Table 3.5). School fees are normally paid in January each year.[13] In 2010, primary school fees on Tubetube were 140 kina per year for Grades 1–5 and 150 kina for Grade 6.[14] Of the 53 households that had children in school, 46 had paid no school fees at all for 2010, while five had paid part, one had paid all and one was unclear.

An interview with the headmaster of the Tubetube primary school revealed that most parents had only paid, on average, a small fraction of their children's school fees even when the bêche-de-mer fishery was still open. The outstanding fees at the end of 2009 were 6,916 kina, which meant an average of 93.46 kina, or two thirds of the amount due for each of the 74 students (see Table 3.6). The headmaster, who was from Ferguson Island, a large island in the northwestern part of Milne Bay Province, observed that: 'Outstanding fees is almost a custom with this group of people. It's different

13   van Helden (2004) argued that the premature opening of the bêche-de-mer season in December 2000 was the result of pressure from villagers needing to pay the next year's school fees.
14   One kina was worth 37 US cents at the time of our fieldwork.

on Ferguson. It never goes over 1,000 kina on Ferguson.' He said that the Tubetube school was able to survive on a small amount of money it received from the national and provincial governments, and they tried to manage the money so they could keep operating until the end of the year.[15]

Table 3.6: Outstanding school fee payments for 2009 at Tubetube primary school

Place of origin of students	Money outstanding (kina)	Number of students
Tubetube	3,255	35
Naluwaluwali	2,453	22
Koyagaugau/Ole	748	12
Anagusa	390	3
Kitai	70	2
TOTAL	6,916	74

Source: Authors' interview data.

The school headmaster on Ware Island corroborated the Tubetube headmaster's story by telling us that the government paid a small subsidy to the school, usually later in the year, to help them keep operating, and had told teachers to allow students to stay in school even if their fees had not been paid. In explaining the high rate of non-payment of fees by Ware parents he referred to the unfortunate timing of the moratorium, shortly before fees were due to be paid, but he also said that Ware people tend to be 'not wise in spending'. The Ware village enumerator told us that the Ware school chairman had told him that most Ware families with children in school 'did not pay a toea' in school fees in 2009, while a small number paid in part and none paid in full. He also told us that the Ware school committee was busy selling clay pots in Alotau at the time of our survey to try to make up the shortfall in fees from 2009. When asked whether Ware villagers requested help with school fee payments from relatives living in urban centres, the enumerator said that about 40 Ware families were in the habit of doing this and typically made visits for this purpose. It is possible that parents were simply exploiting the generosity of the government in allowing their children to stay at school even if fees had not been paid, and we cannot know whether parents would have given a higher priority to school fee payments had the schools prevented students from attending if fees were not paid. But the village enumerator's comment that the committee

---

15 Teachers' salaries were not affected by the deficit since they were paid directly by the government.

was struggling to make up shortfalls by selling clay pots—a highly labour-intensive product—suggests that non-payment of fees was not entirely unproblematic for the schools.

While fees at the small island primary schools were relatively modest in 2010, fees for high school students in the provincial capital, Alotau, or at Bwagoia on Misima Island, reportedly ranged between 1,000 and 2,000 kina. Nevertheless, our overriding impression was that putting aside money for school fees was a not a high priority among most of the people we surveyed and interviewed. Many fishers reported being able to make more than 1,000 kina with a good harvest of bêche-de-mer, but a large proportion of this money was commonly spent on fuel for the next fishing trip, while most of the remainder was spent on food and other store goods. The village enumerator and two other key informants emphasised that lavish expenditure on beer was mainly by young, unmarried men, and the most common expenditure items for most people were food and fuel.[16]

On less densely populated Tubetube and Naluwaluwali, the lack of money following the bêche-de-mer moratorium had less of an impact on food security because people had sufficient fertile land to be able to resume subsistence gardening, even though this was clearly not how they preferred to live. Comments by several survey respondents on Tubetube suggested that many people did less work in their gardens when the fishery was open because they could afford to buy all or most of their food. By contrast, on Ware, our 24-hour diet recall question revealed that a lot of people were only eating one meal a day after the moratorium, and several were resorting to eating the stems of young banana plants. The Ware village enumerator also said that people had been steadily leaving Ware, even before the moratorium, because of declining harvests and the increasing cost of the fuel required to reach unfished sea cucumber stocks. He estimated that the cost of living on Ware was now about 20 kina a day, and without this minimum amount of money, life was quite difficult because of the extreme shortage of land.

The 2011 national census found that the island's resident population had grown to 1,250, or 568 people per square kilometre (see Table 3.2). This meant that the population had grown by more than two thirds since the previous national census in 2000. This rate of growth is only consistent with the village enumerator's observation if we assume that the population

---

16   Hayes (1993) also noted the importance of food, primarily rice and sugar, in his data on Ware Islanders' expenditure patterns in the early 1990s.

was even bigger at the height of the bêche-de-mer boom. Since the average garden fallow period had already been reduced to about two years in 2005, when there were 409 people per square kilometre, it appeared that the population had well and truly overshot the agricultural carrying capacity of the island by 2010, and this was confirmed by the results of our survey of local diets. The village enumerator also noted that land disputes were now very common on the island, as some of the senior islanders had forecast in the early 1990s (Hayes 1993).

The data we present here probably raise more questions than they answer. Because we only became aware of the pattern of non-payment of school fees late in the course of our fieldwork, we did not have time to explore the issue further with more in-depth interviews or surveys. Many interviews with bêche-de-mer fishers, along with several key informant interviews, suggested that most fishers could easily afford to pay the relatively small fees for primary school students while the fishery was still open. One fisher told us he had been able to buy a dinghy, outboard motor and roofing iron with earnings from bêche-de-mer. Would most parents have found the money if schools had taken a harder line with their enrolment policy?

A useful clue to answering this question may lie with the changed pattern of remittance payments back to the islands by employed relatives living in towns. Our survey data and comments by the Ware village enumerator indicate that remittances flowing to the islands were few and small, both before and after the moratorium.[17] This contrasts with the earlier finding by Hayes (1993), who reported that cash remittances accounted for a third of average household incomes on Ware in the early 1990s, and that remittances in kind or 'cargo' were also significant, albeit of lower value than the cash contributions. Perhaps the more recent decline in the value of remittances is sufficient to explain the apparent lack of commitment to the payment of school fees. One possible (but inevitably partial) explanation is the likelihood that the boom in the bêche-de-mer fishery over the preceding two decades created an illusion among islanders that they no longer needed to bother investing in education as a means of long-term social reproduction on the islands.

---

17 A sample survey conducted before the imposition of the moratorium found that wages and remittances from employment outside the agricultural and fisheries sectors accounted for 13 per cent of rural household incomes across Milne Bay Province as a whole (Kaly 2006).

At the same time, the boom appears to have distributed household incomes more widely among the islanders. Two fishers stated that they thought the bêche-de-mer fishery functioned as a great economic equaliser, and they lamented its closure for that reason. On Ware at least, several people commented that the benefits of the bêche-de-mer fishery tended to be shared around generously, so that many older people, particularly widows, also benefited. Although younger men with motor boats and diving gear were able to make more money from their fishing expeditions, this in turn gave rise to complaints that they were violating the egalitarian ethic that applied to the practice of harvesting sea cucumbers in shallow coastal waters without the aid of such technology (Fabinyi et al. 2015). If the benefits of the fishery were widely distributed through the islands, there would be no distinct group of households that got no benefit at all and would therefore have to seek alternative means to improve their livelihoods.

Another Ecosystem Assessment

Although the MBCP was cancelled at the end of 2006 as a result of the mid-term review of its implementation (Baines et al. 2006), Conservation International (CI) did not completely abandon its own plans for the conservation of coastal and marine ecosystems in Milne Bay Province. Instead, they formed a partnership with the Commonwealth Scientific and Industrial Research Organisation (CSIRO), which was able to secure additional funding from the Australian Government's contribution to the Coral Triangle Initiative (CTI)—a regional marine conservation program, with multiple partners and funding sources, which was established in 2009. The overall aim of the CTI was to achieve biodiversity conservation, sustainable fisheries and food security through the establishment of marine protected areas and ecosystem-based approaches to fisheries management in the territorial waters and exclusive economic zones of six countries, one of which was PNG. The fourth goal of the CTI was to promote adaptation planning for small island ecosystems and communities threatened by climate change. The partnership between CI and CSIRO was meant to develop an 'adaptation planning process' for Milne Bay Province that would contribute to the PNG Government's National Plan of Action for the CTI (Butler et al. 2014), although the partnership is not actually mentioned in that national plan (GPNG 2010). The aim was therefore quite similar to that of the SMIP Program that had to be aborted in 2006, but there was a somewhat stronger emphasis on the need to build the capacity of local government planners and managers, rather than members of small island communities, to 'promote sustainable development, decrease poverty and avert disasters' (Skewes et al. 2011: 31).

The CSIRO scientists involved in this exercise adopted a conceptual framework that was very similar to the one adopted by the Millennium Ecosystem Assessment (Skewes et al. 2011: 5). The main difference was the distinction now drawn between 'ecosystem assets' and the services that these things provide to human consumers. Ecosystem assets were divided between marine habitats, terrestrial habitats and 'functional groups', and the latter were divided between 'harvested species' (like sea cucumbers) and 'biodiversity' (as represented by the mix of species associated with coral reefs) (ibid.: 59). The services derived from each asset were divided into four categories—cash income, food, regulating and cultural—and then roughly quantified. By way of example, sea cucumbers scored very highly for income, very low for food and regulation, and had no cultural value at all.

The entire province was then divided into 15 zones or regions that were distinguished by the mix of drivers or threats to which their ecosystem assets and services were subject. The boundaries of these zones bore no particular relationship to the political boundaries between districts or LLG areas. For example, the 'Dawson' zone included the Koyagaugau island community from the Bwanabwana LLG area and a couple of other island communities from the neighbouring Maramatana LLG area, which happen to be those on which CI staff lavished most of their attention between 2004 and 2006.[18]

The drivers or threats were themselves divided between two main categories —climate change and 'human population'. Population pressure was thought to include a number of more specific factors, including resource use, land use, pollution and contaminants. Human population density in each of the 15 zones was calculated separately for land area, coral reef area and sea area. For example, the Dawson zone was said to contain nine distinct islands, inhabited by 2,034 people in 2000, with a combined land area of 19 km² within a total area of 1,027 km². It is not clear how the reef areas and sea areas were allocated to the populations of different island communities or how this allocation might have taken account of the ongoing territorial disputes between them.

---

18 The Maramatana islands are Iabam, Pahilele and Nuakata. The first two belong to a single council ward.

An 'asset-threat interaction matrix' was then constructed for the whole province and each of its 15 zones, with predicted scores for 2030 and 2100. Perhaps not surprisingly, 'human related threats' had more of an impact by 2030, but climate threats had more of an impact by 2100. The assessment predicted that the current high rate of population growth across all 15 zones would decline with higher levels of education or migration levels as local ecosystems and their assets were increasingly over-exploited (Skewes et al. 2011: 98).

The document containing this assessment was a very long one—more than 100 pages—with numerous tables and graphs illustrating the calculations made by the scientists. How would it influence the decisions taken by local government planners and managers? In 2010, CI staff convened a workshop in Alotau, the provincial capital, which was meant to elicit the 'tacit knowledge' of 26 local stakeholders through a 'rapid participatory assessment' (Butler et al. 2014). The participants were apparently drawn from private companies and civil society organisations, as well as from government agencies, and represented five of the 15 zones identified in the ecosystem assessment, including the 'Dawson' zone but not the 'Samarai' zone that included most of the Bwanabwana islands.

A summary version of the ecosystem assessment was presented to the audience by CSIRO staff, and participants from each of the five zones were then asked to score the relative importance of ecosystem services identified in the assessment, to predict likely changes in ecosystem assets by 2030, and then to identify direct and indirect drivers of these changes and possible management strategies to deal with them. They were only asked to consider 'provisioning' services, not regulating or cultural services, because of the risk that they might get confused about the categories. Since they had less than two days in which to reach their conclusions, there was not enough time for 'analysis of social-ecological system thresholds, the agency of decision-makers over drivers, future scenarios and trade-offs' between different services (Butler et al. 2014: 4). Given the simplicity of the task, their conclusions were fairly predictable: the most significant direct driver of change was overfishing, and the most significant indirect driver was population growth. The CSIRO scientists who reported this outcome acknowledged our earlier argument that 'monetisation of the local economy is driving materialism, erosion of traditional norms, institutions and leadership, plus drug, alcohol and debt problems among younger generations, and disputes over land and sea tenure' (ibid.: 7), but these issues were not discussed in the workshop, possibly because the participants were not aware of the operation of these 'community-scale drivers'.

A Broader Context

Despite the substantial investment of time and money in the second ecosystem assessment, there is no evidence that the exercise had a significant impact on decision-making at any level of political organisation. It certainly had less of an impact on the livelihoods of the people living on Milne Bay's numerous small islands than the continuation of the nationwide moratorium on the harvesting of sea cucumbers and production of bêche-de-mer. The moratorium was not lifted until 2017, despite protests from coastal communities around the country, was reimposed again in 2019, lifted again in 2020, and imposed again in 2021.

Jeff Kinch was able to observe its impact from the vantage point of Brooker Island, a very small and densely populated island in the Louisiade LLG area, which he has been visiting on a regular basis since conducting his PhD fieldwork there in the late 1990s (Kinch 2020). Kinch and his colleagues conducted a survey of Brooker and a number of other small islands in the same LLG area at the end of 2014, mainly in order to find out whether the harvest and sale of shark fins had enabled the islanders to compensate for the loss of income from bêche-de-mer (Vieira et al. 2017). They found that this had not happened because the two activities are mutually dependent, which might have been good news for the sharks that are highly vulnerable to this form of exploitation, but was not such good news for the islanders. On Brooker Island, they found that average household cash incomes had fallen by more than 90 per cent in the year following the imposition of the moratorium. The islanders responded to this dramatic fall by planting more food crops, rehabilitating old coconut plantations, engaging in more inter-island trade, making more use of traditional sailing canoes as opposed to motorised dinghies (see Figure 3.5), and spending less money on traditional feasting. When the moratorium was lifted in 2017, they were so keen to make up for lost opportunities that a large number of undersized sea cucumbers were harvested and a large proportion were not properly processed, which meant that the bêche-de-mer went to waste (Kinch 2020).

Figure 3.5: Sailau undergoing renovation on Tubetube Island, 2005
Source: Photograph by Simon Foale.

The moratorium seems to have had a somewhat different impact on the livelihoods and practices of people living on a collection of small islands located between the main island of New Ireland Province and the smaller but still big island of Lavongai (New Hanover). These are generally known as the Tigak islands since this is the name of the language spoken by most of their indigenous inhabitants. Jeff Kinch, Simon Foale and a number of colleagues investigated the impact within the context of a donor-funded project that aimed to encourage the residents of three islands in the group to experiment with the cultivation of the highly valued and hence endangered sandfish species (*Holothuria scabra*) in locally managed marine protected areas (Hair et al. 2016, 2019, 2020; Purdy et al. 2017; Hair 2020). Surveys conducted in 2004 and 2014 found that the moratorium had not resulted in a huge decline in household incomes, as it had on Brooker Island, but that islanders had responded by increasing their harvest of other hand-collected species closer to shore, with women doing more of the work (Purdy et al. 2017). These surveys also found an increase in the proportion of households enforcing communal territorial access rights, although this process of territorialisation might well have occurred without the moratorium, just as it had when the bêche-de-mer fishery was still thriving in Milne Bay.

The lifting of the moratorium in 2017 was based on a revised nationwide management plan that again set a 'total allowable catch' for each province, and allowed for the harvest of sea cucumbers to proceed until this limit was reached or for a maximum period of six months (April to September). As in other parts of the country, New Ireland's limit was reached and exceeded well within the six-month period as the price of bêche-de-mer had risen during the eight years of the moratorium (Hair et al. 2019). The limit was exceeded by an even greater margin and in a shorter period in the 2018 fishing season (Hair et al. 2020). This did not bode well for the sandfish mariculture project as two of the three 'sea ranches' were invaded by poachers, most of whom were members of the island communities that were supposedly responsible for their protection. The failure of island communities to enforce this particular bundle of territorial access rights was ascribed to

> a divided community and fragmented, conflict-ridden local leadership, resulting in a weak communal management system. Overlaying these social and cultural factors, exacerbating external forces were witnessed in the 2018 season including record BDM [bêche-de-mer] prices, strong pressure from buyers, a short open season and absence of control by government regulators.
>
> (Hair et al. 2020: 7)

The apparent failure of the sandfish mariculture project in New Ireland Province, like the earlier failure of the MBCP, raises a broader question about the feasibility of 'community-based' approaches to the conservation of marine biodiversity or the sustainable management of marine resources. Some would argue that there is really no alternative to such approaches so long as the resources in question are subject to customary property rights (Ruddle et al. 1992; Cinner and Aswani 2007; Cinner et al. 2012; Purdy et al. 2017). Others would question whether customary institutions, and the local knowledge systems with which they are associated, have the capacity to produce sustainable outcomes when they are being undermined by the invasion of market forces (Foale and Manele 2004; Foale et al. 2011, 2016; Cohen and Foale 2013; Fabinyi et al. 2015). The commercialisation of marine resources may well encourage coastal communities to make stronger territorial claims and impose stricter controls over access to marine resources, but this process of territorialisation may not be sufficient to diminish the rate at which scarce or endangered resources are exploited by members of these communities (Carrier 1981; Otto 1997; Foale and Macintyre 2000; van Helden 2004; Cinner and Aswani 2007).

The regulations imposed by government authorities, like the bêche-de-mer moratorium, are more effective instruments of conservation (Steenbergen et al. 2019), but in some contexts these may also fail to be effective if local resource owners or alien intruders are able to subvert them, or if the regulating agencies themselves are corrupt or incompetent (Purcell et al. 2013; Eriksson et al. 2015). If the National Fisheries Authority were to cancel the licences of trading companies that purchase undersized bêche-de-mer, then fishers would soon discover that there is no market for this commodity and change their own behaviour accordingly, but this has not yet happened. Overfishing in low-income areas to supply high-value global markets is an example of a wicked problem in fisheries management, which means that no one group of actors can resolve it by themselves, and might fail to do so even if they can find new ways to collaborate with each other (Anderson et al. 2011; Barclay et al. 2019).

Conclusion

The SMIP Program was attached to the MBCP in the expectation or hope that an assessment of the relationship between ecosystem services and human well-being on small, densely populated islands would help decision-makers to make better decisions about the management of scarce resources, and even encourage the islanders themselves to support the establishment of marine protected areas. It is commonly argued that this kind of approach to the management of small-scale artisanal fisheries is one that is likely to be more effective than the imposition of government regulations based on conventional assessments of stock levels because it leads to a higher level of community or stakeholder participation (Allison and Ellis 2001; Purdy et al. 2017). But it is hard to tell whether it has had any positive effect in the Bwanabwana islands or the rest of Milne Bay Province, first because of the demise of the MBCP in 2006 and then because of the imposition of the bêche-de-mer moratorium in 2009. While the moratorium might have made the livelihoods of the islanders more 'sustainable' by limiting the rate at which they were depleting the stock of sea cucumbers, it did nothing to enhance their sense of environmental justice because the stocks were still subject to poaching by Vietnamese 'blue boats' during that period (Song et al. 2019; Kinch 2020) (see Figure 3.6).

Figure 3.6: 'Blue boat' moored at Kavieng Wharf, 2017
Source: Photograph by Simon Foale.

Considerations of justice, fairness or equity are not easily accommodated in an ecosystem assessment for the simple reason that they are much harder to measure than variables like population pressure or the depletion of natural resources (Lau et al. 2020). But if social norms are left out of the equation, it is hard to escape the kind of Malthusian assumptions that seem to have informed the 'rapid participatory assessment' that was conducted in 2010 (Butler et al. 2014). It may well be true that the market value of marine commodities like bêche-de-mer will cause the members of small island communities to deplete the resource whenever they have the opportunity to do so (Purcell et al. 2018). And it is hard to see why they would choose to do otherwise if that is the easiest way for them to make the money they need in order to compensate for a reduction in the supply of other local ecosystem services to a rapidly growing population. But if the production of these commodities is unsustainable in its current form, then something else will sooner or later have to be done.

Participants in the 2010 workshop, who were not members of the small island communities under pressure, thought the only obvious solution to this problem was for more children to get a decent education, find themselves jobs in the formal economy, and possibly keep the islands afloat, in an economic sense, by means of remittances. This certainly appears to have been the strategy adopted by members of several small island communities in Manus Province, where the value of remittances was only slightly lower than the combined value of timber and marine commodity exports in 2006 (Dalsgaard 2013). In the late 1970s, remittances are said to have paid for roughly 80 per cent of what the very small island community of Ponam spent on imported goods (Carrier and Carrier 1989: 167–8). However, relationships between donors and recipients are not without moral hazard, depending as they do on the capacity of migrants to negotiate the moral politics underpinning the demands made of them while they are employed, as well as maintaining adequate levels of knowledge of local kinship networks, land boundaries, customary feasting protocols and the local language.

> In the long run migrants may become too disengaged from village affairs and thus no longer be regarded as taking part in their social reproduction. Then the impulse to educate children for future remittances may disappear.
>
> (Dalsgaard 2013: 299)

The combination of low *voluntary* remittance flows to the Bwanabwana islands with the more explicit demands on urban relatives for assistance with school fees, as reported by the Ware village enumerator in 2010, may indicate a higher level of disengagement, and even the espousal of an ideology of 'possessive individualism', on the part of the urban relatives (Martin 2007). But the earlier bêche-de-mer boom seems to have reduced the demand, as well as the supply, and may well have been more egalitarian in its effects on local household incomes. It is hard to tell whether the subsequent closure of the fishery served to magnify tensions in the relationship between the islanders and their absent relatives, or whether these tensions were moderated by the national government's adoption of a free education policy in 2012.[19] But we need to be wary of assuming that the pattern of migration and remittances fits neatly within a dualistic conception of the relationship between 'modern' (urban) and 'traditional' (rural) households, institutions or practices.

---

19 This policy was not implemented consistently across the country, and was partially abandoned in 2019, so most parents still had to contribute something towards the cost of their children's education, even if the amount varied from place to place or year to year (Walton and Hushang 2021).

As Hayes remarked in 1993, there is more than one circuit or form of circulation whereby people, goods and 'ecosystem services' move across the boundary that separates a small island community from the rest of the world, and these boundary conditions can change in response to numerous factors or drivers aside from market prices or population pressure. This is evident in the way that Brooker Islanders, and no doubt other islanders, responded to the imposition of the moratorium. Our 2006 survey data also revealed the extent to which islanders in the Bwanabwana group have actively sought to maintain the option of moving from one island to another, or one community to another, even when they do not have the option, or even the desire, to migrate to an urban centre. There is strong evidence that this livelihood strategy evolved over many centuries, primarily as a response to the risk of periodic food shortages induced by climatic fluctuations (Macintyre 1983; Macintyre and Allen 1990). But we should not therefore conclude that its continued viability depends on the maintenance of customary institutions that are now being eroded by the forces of modernity.

References

Allen, G.R., J.P. Kinch, S.A. McKenna and P. Seeto, 2003. *A Rapid Marine Biodiversity Survey of Milne Bay Province: Survey II (2000)*. Washington DC: Conservation International (RAP Bulletin of Biological Assessment 29).

Allison, E. and F. Ellis, 2001. 'The Livelihoods Approach and Management of Small-Scale Fisheries.' *Marine Policy* 25: 377–388. doi.org/10.1016/S0308-597X (01)00023-9

Anderson, S., J. Flemming, R. Watson and H. Lotze, 2011. 'Serial Exploitation of Global Sea Cucumber Fisheries.' *Fish and Fisheries* 12: 317–339. doi.org/ 10.1111/j.1467-2979.2010.00397.x

Baines, G., J. Duguman and P. Johnston, 2006. 'Milne Bay Community-Based Coastal and Marine Conservation Project: Terminal Evaluation of Phase 1.' Port Moresby: United Nations Development Programme.

Balboa, C.M., 2013. 'How Successful Transnational Non-Governmental Organizations Set Themselves up for Failure on the Ground.' *World Development* 54: 273–287. doi.org/10.1016/j.worlddev.2013.09.001

Barclay, K., M. Fabinyi, J. Kinch and S. Foale, 2019. 'Governability of High-Value Fisheries in Low-Income Contexts: A Case Study of the Sea Cucumber Fishery in Papua New Guinea.' *Human Ecology* 47: 381–396. doi.org/10.1007/s10745-019-00078-8

Belshaw, C.S., 1955. *In Search of Wealth: A Study of Emergent Commercial Operations in the Melanesian Society of Southeastern Papua.* Memoirs of the American Anthropological Association (Memoir 80).

Butler, J.R.A., T. Skewes, D. Mitchell and others, 2014. 'Stakeholder Perceptions of Ecosystem Service Declines in Milne Bay, Papua New Guinea: Is Human Population a More Critical Driver Than Climate Change?' *Marine Policy* 46: 1–13. doi.org/10.1016/j.marpol.2013.12.011

Carrier, J.G., 1981. 'Ownership of Productive Resources on Ponam Island, Manus Province.' *Journal de la Société des Océanistes* 37: 205–217. doi.org/10.3406/jso.1981.3061

Carrier, J.G. and A.H. Carrier, 1989. *Wage, Trade and Exchange in Melanesia.* Berkeley: University of California Press.

Cinner, J.E. and S. Aswani, 2007. 'Integrating Customary Management into Marine Conservation.' *Biological Conservation* 140: 201–216. doi.org/10.1016/j.biocon.2007.08.008

Cinner, J.E., T.R. McClanahan, M.A. MacNeil and others, 2012. 'Comanagement of Coral Reef Social-Ecological Systems.' *Proceedings of the National Academy of Sciences* 109: 5219–5222. doi.org/10.1073/pnas.1121215109

Cohen, P.J. and S.J. Foale, 2013. 'Sustaining Small-Scale Fisheries with Periodically Harvested Marine Reserves.' *Marine Policy* 37: 278–287. doi.org/10.1016/j.marpol.2012.05.010

Dalsgaard, S., 2013. 'The Politics of Remittance and the Role of Returning Migrants: Localizing Capitalism in Manus Province, Papua New Guinea.' *Research in Economic Anthropology* 33: 277–302. doi.org/10.1108/S0190-1281(2013)0000033013

Eriksson, H., H. Österblom, B. Crona and others, 2015. 'Contagious Exploitation of Marine Resources.' *Frontiers in Ecology and the Environment* 13: 435–440. doi.org/10.1890/140312

Fabinyi, M., S. Foale and M. Macintyre, 2015. 'Managing Inequality or Managing Stocks? An Ethnographic Perspective on the Governance of Small-Scale Fisheries.' *Fish and Fisheries* 16: 471–485. doi.org/10.1111/faf.12069

Filer, C., 2002. 'Small Islands in Peril in Milne Bay Province.' Canberra: The Australian National University, Research School of Pacific and Asian Studies, Resource Management in Asia-Pacific Program (unpublished submission to United Nations Development Programme).

——, 2009. 'A Bridge Too Far: The Knowledge Problem in the Millennium Assessment.' In J.G. Carrier and P. West (eds), *Virtualism, Governance and Practice: Vision and Execution in Environmental Conservation*. New York: Berghahn Books.

Foale, S., 2005. 'Sharks, Sea Slugs and Skirmishes: Managing Marine and Agricultural Resources on Small, Overpopulated Islands in Milne Bay, PNG.' Canberra: The Australian National University, Research School of Pacific and Asian Studies, Resource Management in Asia-Pacific Program (Working Paper 64).

Foale, S., P. Cohen, S. Januchowski and others, 2011. 'Tenure and Taboos: Origins and Implications for Fisheries in the Pacific.' *Fish and Fisheries* 12: 357–369. doi.org/10.1111/j.1467-2979.2010.00395.x

Foale, S., M. Dyer and J. Kinch, 2016. 'The Value of Tropical Biodiversity in Rural Melanesia.' *Valuation Studies* 4: 11–39. doi.org/10.3384/VS.2001-5992. 164111

Foale, S. and M. Macintyre, 2000. 'Dynamic and Flexible Aspects of Land and Marine Tenure at West Nggela: Implications for Marine Resource Management.' *Oceania* 71: 30–45. doi.org/10.1002/j.1834-4461.2000.tb02722.x

Foale, S. and B. Manele, 2004. 'Social and Political Barriers to the Use of Marine Protected Areas for Conservation and Fishery Management in Melanesia.' *Asia Pacific Viewpoint* 45: 373–386. doi.org/10.1111/j.1467-8373.2004.00247.x

GPNG (Government of Papua New Guinea), 2010. 'PNG Marine Program on Coral Reefs, Fisheries and Food Security 2010–2013: National Plan of Action on the Coral Triangle Initiative.' Port Moresby: Department of Environment and Conservation and National Fisheries Authority.

GPNG (Government of Papua New Guinea) and UNDP (United Nations Development Programme), n.d. [2005]. 'Community-Based Coastal and Marine Conservation in Milne Bay Province.' Port Moresby: GPNG and UNDP (unpublished project document).

Hair, C., 2020. Development of Community-Based Mariculture of Sandfish, *Holothuria scabra*, in New Ireland Province, Papua New Guinea. Townsville: James Cook University (PhD thesis).

Hair, C., S. Foale, N. Daniels and others, 2020. 'Social and Economic Challenges to Community-Based Sea Cucumber Mariculture Development in New Ireland Province, Papua New Guinea.' *Marine Policy* 117: 103940. doi.org/10.1016/j.marpol.2020.103940

Hair, C., S. Foale, J. Kinch and others, 2016. 'Beyond Boom, Bust and Ban: The Sandfish (*Holothuria scabra*) Fishery in the Tigak Islands, Papua New Guinea.' *Regional Studies in Marine Science* 5: 69–79. doi.org/10.1016/j.rsma.2016.02.001

——, 2019. 'Socioeconomic Impacts of a Sea Cucumber Fishery in Papua New Guinea: Is There an Opportunity for Mariculture?' *Ocean and Coastal Management* 179: 104826. doi.org/10.1016/j.ocecoaman.2019.104826

Hayes, G., 1993. '"MIRAB" Processes and Development on Small Pacific Islands: A Case Study from the Southern Massim, Papua New Guinea.' *Pacific Viewpoint* 34: 153–178. doi.org/10.1111/apv.342002

Hide, R.L., R.M. Bourke, B.J. Allen and others, 2002. 'Milne Bay Province: Text Summaries, Maps, Code Lists and Village Identification.' Canberra: The Australian National University, Department of Human Geography (Agricultural Systems of Papua New Guinea Working Paper 6).

Kaly, U., 2006. 'Socio-Economic Survey of Small Scale Fisheries in Milne Bay Province, Papua New Guinea.' Kavieng: National Fisheries Authority and Coastal Fisheries Management and Development Project.

Kinch, J., 2001. 'Social Evaluation Study for the Milne Bay Community-Based Coastal and Marine Conservation Program.' Port Moresby: Conservation International (unpublished report to the United Nations Development Programme).

——, 2002. 'Overview of the Beche-de-Mer Fishery in Milne Bay Province, Papua New Guinea.' Noumea: SPC (Beche-de-Mer Information Bulletin 17).

——, 2007. 'A Review of Fisheries and Marine Resources in Milne Bay Province, Papua New Guinea.' Kavieng: National Fisheries Authority and Coastal Fisheries Management and Development Project.

——, 2020. Changing Lives and Livelihoods: Culture, Capitalism and Contestation over Marine Resources in Island Melanesia. Canberra: The Australian National University (PhD thesis).

Kinch, J., S. Purcell, S. Uthicke and K. Friedman, 2008. 'Papua New Guinea: A Hotspot of Sea Cucumber Fisheries in the Western Central Pacific.' In V. Toral-Granda, A. Lovatelli and M. Vasconellos (eds), *Sea Cucumbers: A Global Review of Fisheries and Trade*. Rome: UN Food and Agriculture Organization (Fisheries and Aquaculture Technical Paper 516).

Lau, J.D., J.E. Cinner, M. Fabinyi and others, 2020. 'Access to Marine Ecosystem Services: Examining Entanglement and Legitimacy in Customary Institutions.' *World Development* 126: 104730. doi.org/10.1016/j.worlddev.2019.104730

MA (Millennium Ecosystem Assessment), 2003. *Ecosystems and Human Well-Being: A Framework for Assessment.* Washington DC: Island Press.

Macintyre, M., 1983. Changing Paths: An Historical Ethnography of the Traders of Tubetube. Canberra: The Australian National University (PhD thesis).

——, 1987. 'Nurturance and Nutrition: Change and Continuity in Concepts of Food and Feasting in a Southern Massim Community.' *Journal de la Société des Océanistes* 84: 51–59. doi.org/10.3406/jso.1987.2561

——, 1989. 'Better Homes and Gardens.' In M. Jolly and M. Macintyre (eds), *Family and Gender in the Pacific: Domestic Contradictions and the Colonial Impact.* Melbourne: Cambridge University Press. doi.org/10.1017/CBO9781 139084864.009

Macintyre, M. and J. Allen, 1990. 'Trading for Subsistence: The Case from the Southern Massim.' In D.E. Yen and J.M.J. Mummery (eds), *Pacific Production Systems: Approaches to Economic Prehistory.* Canberra: The Australian National University, Research School of Pacific Studies, Department of Prehistory (Occasional Paper 18).

Macintyre, M. and M. Young, 1982. 'The Persistence of Traditional Trade and Ceremonial Exchange in the Massim.' In R.J. May and H. Nelson (eds), *Melanesia: Beyond Diversity.* Canberra: The Australian National University, Research School of Pacific Studies.

Martin, K., 2007. 'Your Own *Buai* You Must Buy: The Ideology of Possessive Individualism in Papua New Guinea.' *Anthropological Forum* 17: 285–298. doi.org/10.1080/00664670701637743

McAlpine, J. and G. Keig, 1983. *Climate of Papua New Guinea.* Canberra: Commonwealth Scientific and Industrial Research Organisation and Australian National University Press.

Mitchell, D.K., J. Peters, J. Cannon and others, 2001. 'Sustainable Use Option Plan for the Milne Bay Community-Based Coastal and Marine Conservation Program.' Unpublished report to Conservation International.

Otto, T., 1997. 'Baitfish Royalties and Customary Marine Tenure in Manus, Papua New Guinea.' *Anthropological Forum* 7: 667–690. doi.org/10.1080/00664677. 1997.9967479

Purcell, S., A. Mercier, C. Conand and others, 2013. 'Sea Cucumber Fisheries: Global Analysis of Stocks, Management Measures and Drivers of Overfishing.' *Fish and Fisheries* 14: 34–59. doi.org/10.1111/j.1467-2979.2011.00443.x

Purcell, S.W., D.H. Williamson and P. Ngaluafe, 2018. 'Chinese Market Prices of Beche-de-Mer: Implications for Fisheries and Aquaculture.' *Marine Policy* 91: 58–65. doi.org/10.1016/j.marpol.2018.02.005

Purdy, D.H., D.J. Hadley, J.O. Kenter and J. Kinch, 2017. 'Sea Cucumber Moratorium and Livelihood Diversity in Papua New Guinea.' *Coastal Management* 45: 161–177. doi.org/10.1080/08920753.2017.1278147

Ruddle, K., E. Hviding and R. Johannes, 1992. 'Marine Resources Management in the Context of Customary Tenure.' *Marine Resource Economics* 7: 249–273. doi.org/10.1086/mre.7.4.42629038

Russell, P., 1970. 'The Papuan Beche-de-Mer Trade to 1900.' Waigani: University of Papua New Guinea (MA thesis).

Sabetian, A. and S. Foale, 2006. 'Evolution of the Artisanal Fisher.' *Traditional Marine Resource Management Bulletin* 20: 3–10.

Scales, H., A. Balmford, M. Liu and others, 2006. 'Keeping Bandits at Bay?' *Science* 313: 612–613. doi.org/10.1126/science.313.5787.612c

Skewes, T., J. Kinch, P. Polon and others, 2002. 'Research for Sustainable Use of Beche-de-Mer Resources in Milne Bay Province, Papua New Guinea.' Cleveland: CSIRO Division of Marine Research.

Skewes, T., V. Lyne, J. Butler and others, 2011. 'Melanesian Coastal and Marine Ecosystem Assets: Assessment Framework and Milne Bay Case Study.' CSIRO Final Report to the CSIRO AusAID Alliance.

Smaalders, M. and J. Kinch, 2003. 'Canoes, Subsistence and Conservation in the Louisiade Archipelago of Papua New Guinea.' *Traditional Marine Resource Management Bulletin* 15: 11–20.

Song, A., V. Hoang, P. Cohen and others, 2019. '"Blue Boats" and "Reef Robbers": A New Maritime Security Threat for the Asia Pacific?' *Asia Pacific Viewpoint* 60: 310–324. doi.org/10.1111/apv.12240

Steenbergen, D.J., M. Fabinyi, K. Barclay and others, 2019. 'Governance Interactions in Small-Scale Fisheries Market Chains: Examples from the Asia-Pacific.' *Fish and Fisheries* 20: 697–714. doi.org/10.1111/faf.12370 van Helden, F., 2004. '"Making Do": Integrating Ecological and Societal Considerations for Marine Conservation in a Situation of Indigenous Resource Tenure.' In L.E. Visser (ed.), *Challenging Coasts: Transdisciplinary Excursions into Integrated Coastal Zone Development.* Amsterdam: Amsterdam University Press. doi.org/10.1017/9789048505319.006

Vieira, S., J. Kinch, W. White and L. Yaman, 2017. 'Artisanal Shark Fishing in the Louisiade Archipelago, Papua New Guinea: Socio-Economic Characteristics and Management Options.' *Ocean & Coastal Management* 137: 43–56. doi.org/10.1016/j.ocecoaman.2016.12.009

Walton, W.G. and H. Hushang, 2021. 'The Politics of Undermining National Fee-Free Education Policy: Insights from Papua New Guinea.' *Asia & the Pacific Policy Studies* 8: 401–419. doi.org/10.1002/app5.339

4

Pilot Fish Rock, or, How to Live Large on a Small Island in Marovo Lagoon, Solomon Islands

Edvard Hviding

A Large, Small Island

In this chapter I address a particular question raised in the Introduction, which involves the extent to which islands are not just islands but also nodes in economic, social and political networks. I do this by focussing on a rather small island that has a very central location and position—geographically, historically, politically, economically—in the large lagoon of Marovo in the western Solomon Islands. The island discussed in this chapter is by some definitions small, but in terms of its multitude of relationships and their spatial extent, that small island is big indeed. The Marovo Lagoon itself is famous as a troubled biodiversity hotspot, in which agents of global conservation have competed for attention and influence with multinational logging companies over several decades, and in which some very unique configurations of local politics and economy have developed in response to diverse external cross-pressures (Hviding and Bayliss-Smith 2000; Hviding 2003a, 2006, 2011, 2015a; Duke et. al. 2007). What is less well known from this complex and interesting scene of the present is the particular role played by the small island of Tusu Marovo (literally 'Marovo Island') in the great lagoon's long-term history and political dynamics. It is somewhat notable that on this island neither conservationists nor loggers ever gained substantial footholds, and as for the interests of the latter, Tusu Marovo anyhow has very few trees of commercial value. However, echoing a history of active external engagements, such as enduring interactions with 'Sydney traders' and other European trading vessels that were reported ever since the first nineteenth-century arrivals of ships in the lagoon (Findlay 1877: 773), the people of Tusu Marovo have since the 1980s welcomed a series of collaborative research efforts, some of long duration (Hviding 1996; Duke et. al. 2007). The island has been my own main place of residence throughout several years of fieldwork beginning in 1986 (Hviding 2012). I have been graciously welcomed by the people of Tusu Marovo, who have fed me, housed me, educated me and incorporated me into their social lives. It is from this vantage point that I write, while also referring to published indigenous views of this remarkable island and its history by my close friend and age mate Wilson Liligeto (2006), who researched, wrote and published a book on his own island's history in dialogue with elders, and in conjunction with his career as a senior civil servant.

Tusu Marovo is an entity of several names and several distinct roles, in local and foreign histories. The initial, logical observation is that—as for quite a few other ostensibly small islands in the Pacific—Tusu Marovo is much, much more than its small land mass would indicate. The island is small only in a topographical sense. As with other small islands of its kind, the sociality, economy and history of Tusu Marovo are far bigger than its land. One cannot even attempt to understand the large and complex socio-political scale of Marovo Lagoon without some knowledge of tiny Tusu Marovo. This observation is not confined to Marovo Lagoon; the account I provide here of Tusu Marovo and its place in the world applies to a considerable degree also to the only slightly larger island of Nusa Roviana, located at the far side of New Georgia from Marovo. Nusa Roviana or 'Roviana Island' mirrors Tusu Marovo in being emblematic of a large lagoon area and of a distinctive, linguistically defined district with a history of powerful regional influence (see, for example, Walter and Sheppard 2000, 2017). The significant regional roles of those two equally 'small' islands have their foundations in the fact that in the history of the Marovo and Roviana lagoons it is not the land but the sea that has been—and still is—the scene for mobility, interaction and expansion. Today, the seashores, lagoons, seaways and ocean that were once realms of overseas trading, raiding and headhunting, in and outwards from the western Solomon Islands, constitute a large-scale field of regional logistics, inter-island marriage and long-term social and political alliances (Hviding 1996, 2014a, 2015b). Throughout these stages of history and transformations of society, polity and economy, the small island of Tusu Marovo has remained big, and everyday life on the island has retained a somewhat large-scale quality.

The Making of Pilot Fish Rock and Tusu Marovo

In his book about Tusu Marovo and its people, indigenous historian Wilson Liligeto (2006: 1) writes of his home:

> Tusu Marovo, or Marovo Island, is a tiny piece of mostly hilly land … Despite the island's small size and very limited arable land, it is in many ways the focal point of the entire Marovo Lagoon region.

In the colonial and mission histories of Marovo Lagoon, Tusu Marovo has the distinction of being the location of the first regular trading spot for European ships and of the first Seventh-Day Adventist church, as well as being the home of the first Bible translators of Marovo.

How did a small island become such a big place? As I have already commented, while much has been written about the Marovo Lagoon as a contested wonder of global biodiversity, its homonymous small island is not well known beyond local appreciations such as Liligeto's. From having lived on Tusu Marovo through most of my total of 43 months of fieldwork in Marovo Lagoon, I shall begin to sketch the long-term trajectories of environment, population and political dynamics on and of the island, thereby to disentangle some connections and complexities of this small volcanic peak that has such a central place in a much larger world. Figure 4.1 shows the location of the island at a central point in the lagoon, facing the raised barrier reef that constitutes its outer boundary.

Figure 4.1: Map of Marovo Lagoon showing districts and a selection of main villages
Source: University of Bergen Cartography.

There was a time when there was no 'Marovo Island'. The characteristic peaked topographic silhouette so visible from afar to anyone travelling the wide central lagoon was known instead as Patu Laiti or 'Pilot Fish Rock'. This triangle-shaped island with sections of narrow outlying land measures not much more than 7 km² altogether. It is separated from the large mountainous island of Vangunu by a narrow, mangrove-fringed and shallow but largely navigable channel named Gaio (see Figure 4.2). Yet it is regarded as a totally separate entity from the main island. It falls into the vernacular category of *tusu*, which means 'island' relative to a land mass such as Vangunu, which is classified as *soloso*—a particular Marovo conceptualisation of territorially large-scale dimension that refers to continental islands with tall mountains, or even 'the world'.

Figure 4.2: Tusu Marovo and its associated lagoon and barrier reef from the air, facing due north from above the Vangunu mainland, 2010
Source: Photograph by Edvard Hviding.

This particular position of Tusu Marovo is the source of its original name—it is situated just in front of the 'head' of Vangunu, and is in that regard said to be 'just like the pilot fish that swims just ahead of, and with, the shark'. A diversion into maritime imagery is required at this stage. The close association of pilot fish (the small trevally *Naucrates ductor*) with sharks would seem to be well known in the traditions of tropical coastal peoples and in maritime popular culture but insufficiently studied by marine science. A literature search yields a few not exactly up-to-date published references on aspects of this inter-species symbiotic association, with an emphasis more on sharks than on their small companions (e.g. Hubbs 1951).[1] Interestingly, the Marovo fish taxon of *laiti* is pluralistic and covers several very different fish species, including (1) the free-swimming 'pilot fish' proper, as well as (2) remora sucker-fishes that do not swim ahead of sharks but attach to their bodies and 'hitchhike', and (3) the small cleaner wrasses known for cleaning the mouths of any type of bigger fish on the reef. The conversational context of the usage of the name *laiti* defines which type of fish is the actual point of reference.

---

1   But see Magnuson and Gooding (1971: 315) for interesting behavioural field observations of pilot fish that 'apparently defend the shark as moving territory'.

Figure 4.3: Morning view from the east of the characteristic peaked silhouette of Tusu Marovo, with the barrier islands of central Marovo Lagoon in the background, 2012

Source: Photograph by Edvard Hviding.

The maritime lifestyle of Marovo people is reflected in extraordinarily rich, empirically based local knowledge of fish and other marine organisms—not simply focussed on how to catch them, but including a wealth of minute detail on relationships among species as observed on the reef and at sea (Johannes and Hviding 2000; Hviding 2005). The diverse imagery of *laiti*—an assemblage of dissimilar fish species—draws on the habit of them all to exist in enduring, symbiotic associations with far bigger, and dangerous, creatures—presumably not unlike the way in which the 'rock' now known as Tusu Marovo stands in proximity to the *soloso* ('mainland world') of Vangunu.

The island is obviously the result of violent volcanic activity and has the look of land that has emerged from an eruption. Parts of its terrain rise very steeply. From the rocky shore of the village of Chea, as well as from the flat tidal beach of the nearby village of Chubikopi, a mere 600 m measured point-to-point along the coast rises up to more than 150 m above sea level (see Figure 4.3). Along the northern, lagoon-facing coast are places where hot sulphur-smelling steam and water at boiling point emerge from fissures and cracks in the beach or at a few metres' depth on the lagoon floor itself. In the lower hills are patches of hot sandy soil where thermal forces offer hospitable conditions to the burrowing Melanesian scrub fowl or 'megapode' (*Megapodius eremita*), whose eggs hatch not through the effort of the nesting bird but from the warmth of the sand.

The rugged terrain of the island is subsumed in its most spectacular feature, which constitutes its defining topographical index and can be viewed from afar: Toa Marovo, the steep hill that rises sharply from the northeastern side to a summit of more than 150 m. This sharp ridge of the island's summit has twin adjacent peaks, referred to as Toa Gete ('big peak') and Toa Kiki ('small peak'). It effectively separates the island into the two territorial divisions of Kalelupa ('ocean-side', the northern shore) and Kalekogu ('lagoon-side', the eastern and southeastern shores). These two geographical designations also apply to the two subdivisions of the island's primary resident cognatic descent group Butubutu Marovo, following the pervasive New Georgian pattern whereby a *butubutu* and its ancestral territory or *puava*—an assemblage of land, reefs and sea—are mutually constitutive and share a common name (Hviding 1996, 2003b).

Figure 4.4: Recent aerial view of Tusu Marovo, showing locations of three major villages
Source: Google Earth 2023.

The land of the Butubutu Kalelupa subdivision is densely populated, with two major villages (Sasaghana and Chea) having a total population of more than 800 in 2012, and is under intensive hillside agroforestry. The somewhat larger land of the Butubutu Kalekogu subdivision has one village (Chubikopi), with a population of about 350, and substantial uncultivated secondary forest facing the mainland of Vangunu. Figure 4.4 shows Sasaghana village at the northwestern corner of the island, Chea village at the northeastern corner and Chubikopi village on the southeastern shore. The large primary school at Hinakole, which serves the villages of Chea and Sasaghana, is located on the northern shore, between these two villages.[2] Several small coastal hamlets are located between Chea and Sasagjana.

In agricultural terms, the densely populated Kalelupa land is notable for the extraordinary qualities of its deep, black volcanic soil, which permits continuous cultivation with only minimal periods of fallow. During my first fieldwork on Tusu Marovo in 1986–87, a number of hillside gardens on the north-facing Kalelupa slopes had been under cultivation since they were first established about 1930. The few fallow periods had been so short as to only allow for thickets of easily removable rhizomic gingers to grow. Interspersed with these long-term garden areas are old groves of *Canarium* nut trees, cared for through the generations, and with their white bark highly visible against the dark secondary forest. This is an entirely anthropogenic landscape, transformed over hundreds of years. On the slopes of the Kalekogu area, however, facing the mainland of Vangunu, much less fertile red soils require the normal fallow periods characteristic of Melanesian shifting cultivation, and at no time have the sizeable tracts of forest there been entirely cleared for agriculture. Continued beliefs in malevolent spiritual forces resident in the hills facing the Vangunu mainland make for additional avoidance, and for a forest much less modified by human activity.

Oral Ancient History

To illuminate how Patu Laiti became Tusu Marovo, I now draw on oral traditions of the area and describe some history of place and population from the example of the Kalelupa lands. Genealogies of resident chiefs place the time of permanent settlement by the ancestors of those who live on the island now to about 250–300 years BP, in other words to the early

---

2   The school's recently cleared food gardens are also visible in this image.

eighteenth century, perhaps around 1720 taking conventional genealogical depth into account. However, on the upper slopes of the Marovo peak is a remarkable collection of large, carved and smoothly shaped basalt stone columns 2–3 m long, found in various stages of completion at a location below the summit called Toa Kia ('stonework peak'). These large stone artefacts are considered locally as a firm indication that 'other people', in no way connected to the cultural history of the present population, lived and worked there in a remote past. No one knows what happened to the producers of those large basalt columns, or who they were. However, it is noted that the stone once used to make them was clearly imported from quarries on northern Vangunu, as demonstrated by a number of unfinished columns still visible in various shallow locations on the nearby lagoon floor. Evidently, the transport of the heavy, roughly cut stone columns was arduous, and it is surmised that those early stone-working inhabitants had no more than flimsy rafts at their disposal. People on today's Tusu Marovo express admiration and wonder at the abilities of ancient 'stone age' people to extract the rock, cut the large pieces into shape, transport them by sea and ultimately haul them up to the high slopes for finishing—and, presumably, for placement in a ceremonially significant structure no longer evident. It seems likely that earthquakes have caused any structured assemblage of the large carved columns to tumble long ago.

Figure 4.5: *Buli Te Lagiti* **(Lagiti's petrified throwing clubs), 1996**
Source: Photograph by Edvard Hviding.

Story telling about the remote past of Tusu Marovo and its vicinity make much of a giant ogre by the name of Lagiti, who spent much of his time on the island then, although keeping his home over in the mainland Vangunu hills across from the Gaio channel. In these well known tales Lagiti is portrayed as the powerful maker of significant elements of the island's topography, as well as of the entire barrier reef that defines the Marovo Lagoon itself. The large basalt columns seen at Toa Kia are in fact said to be petrified specimens of Lagiti's repertoire of wooden throwing sticks (*hae buli*) (see Figure 4.5). In the past, carefully balanced sticks of hardwood or thick ginger stems, sharpened at each end, were an efficient technology for capturing flying birds and fruit bats. As for the big stone versions up near the Tusu Marovo peak, their size of course corresponded to that of the giant ogre. They are reckoned to have ended up in disorderly fashion in their present locations after a series of unsuccessful attempts by Lagiti to hit a mythical fruit bat named Vekuveku, which was in the habit of flying over the island's summit on its way to the faraway island of Borokua (an extinct crater in the open ocean to the east), to supply lonely settlers there with significant things not found at Borokua, including such essentials as water, *Canarium* nuts and taro (Hviding 1995: 2–25). Lagiti the giant desired those essentials too and threw many of his giant ginger stem sticks at the bat as it flew across the island, but the summit obstructed his view. Lagiti stepped on the Toa Gete to push it aside for a better view, thereby causing the peak's somewhat precarious sideways tilt, as seen today.

As it was, none of the sticks Lagiti threw at Vekuveku ever succeeded in shooting down the bat, and some crashed against the island's hillsides— where they remain in petrified condition at Toa Kia. But there were some non-lethal hits which caused items carried by Vekuveku to drop to the ground. Through those hits and Vekuveku's loss of what was carried, the much-prized *Canarium* nuts now so abundant on the slopes, as well as fresh water which the bat carried in a leaf, arrived at what had until then been an infertile, waterless island, thereby making the place inhabitable for humans and ogres alike. It should be noted, however, that until a major piped freshwater supply from the mainland rivers was established in the early years of the twenty-first century, water on Tusu Marovo remained a precarious resource, limited to a few natural freshwater springs. Despite Lagiti's efforts, water actually emerged on the island only to a very limited degree.

In due course Lagiti moved over from the Vangunu mainland and settled on the island with his wife. She soon complained that the constant noise made by the surf as it crashed onto the ocean-facing shores (later to become Kalelupa lands) disturbed her sleep at night. There was no lagoon then, but Lagiti grabbed hold of the wave-battered beach, as well as the entire northern shores of Vangunu, and dragged and pushed it outwards way into the ocean to block the ocean surf forever in the form of a raised barrier reef (Hviding 1996: 35). So Lagiti became the maker of that huge lagoon that has ever since offered its human inhabitants so many benefits. The petrified skull of his wife can be seen today in the form of a large, strangely shaped volcanic boulder in the tidal zone between the present village of Chea and the old headhunting stronghold of Babata. And so the small, as yet unnamed, island with the steep summit became the vantage point from which Lagiti's miracles of creative agency had generated the conditions for a distinctive way of human life.[3]

The Lagiti stories are about as far as tales of the island's remote past go, although the tales are not so pervasive as to inhibit ongoing local speculations about who the stoneworkers of Toa Kia may have been and where they came from. In any event, those people were long gone when known human settlement started. It is narrated in the traditions of the Kalelupa people that Kelo, the first chief born on the island after his father Tutikavo had migrated from the Podokana district on southern New Georgia with just a few followers and their families, felt sad for being lonely on that thinly populated island they had named Patu Laiti. One day he exclaimed, '*Ie ma rovo nia he kisa*' ('This [place] makes us lonely').[4] Kelo's complaint appears to have become a well-known standard expression and, having heard it, chiefs who visited from Kelo's home districts said, 'If that is so, then from now on the name of your island is Marovo, not Patu Laiti'.

In Kelo's time, it is said that the small number of people resident on the island had no language of their own, but spoke a mixed vernacular based on the language of the Vangunu mainland, with influences from the northern languages of Hoava and Vahole, whose speakers were frequent visitors to

---

3   I have had the opportunity to discuss this tale with geologists with some knowledge of the Marovo Lagoon, and they have noted that the account of Lagiti's hard work in relocating the seashore to form the raised barrier reef is in fact quite an apt summary of how violent seismic forces of rupture, subsidence and warping across fault lines may have generated the special topography of the lagoon (see also Stoddart 1965).

4   *Ma* is the first person exclusive pronoun that here links to the adjective *rovo*, which refers to a particular condition of experienced social isolation.

the central lagoon. However, following extended contact with visitors from the Kalivarana (Viru Harbour) area of southern New Georgia, where the founding ancestor Tutikavo had connections, the increasing population of the now renamed Tusu Marovo felt a need for a language of their own. Hence, it is recounted how they purchased the Kalivarana language—in other words, obtained the permanent right to speak it. The Tusu Marovo people have since used and refined it over the generations, to the degree that the island today is considered to be the 'true' repository of *jinama Marovo*, ('Marovo language'), although the language is referred to in the Roviana area as *zinama Ulusaghe* (a regional designation), and is still known in its original location as *jinama Kalivarana*.

This account of the sketchy, piecemeal coming-into-being of a Tusu Marovo that was to become a formidable power on an inter-island scale, and of the name 'Marovo' itself, is interesting background for what was to come. The traders who had their first friendly encounters with the people of the island sometime around the mid-nineteenth century appropriated its name to apply to the entire region of the vast lagoon that was gradually discovered and explored by them—a charting process that was later followed up by Royal Navy surveyors (Findlay 1877: 773; Somerville 1897: 360). Once the somewhat humorously invented name of a small island, 'Marovo' then became the designation applied by European navigators to whole of the southeastern part of the New Georgia archipelago that was defined by the lagoon, and that corresponded quite well with the local district name of Ulusaghe (Hviding 1996: 34, 104–6). This exogenous expansion of 'Marovo' must be seen, however, as a reflection of the rather formidable political and military strength that had been built on Tusu Marovo long before the first European traders ventured into the lagoon.

'Eyes That Could See Everywhere'

From Kelo's time as chief, strongholds of warriors with regional power started to develop on the island. War canoes from Tusu Marovo soon ranged far and wide in the seasonal quest for heads, which had a climax during the long periods of calm weather around what are now the months of October and November, between the seasons of dry southeast trade winds and wet monsoons from northwest. Tusu Marovo's warriors became a particular scourge of the people in the Bughotu area of southern Isabel, almost 100 km across the open sea, and some canoe crews paddled as far as the distant lands of Visale and Talise in what is now Guadalcanal, about 200 km from Marovo. A diversity of oral traditions gives details on this extraordinary, predatory maritime mobility. Journeys to Tete Ulu, the old Marovo name for western Guadalcanal, allowed for convenient stopovers along the way at the uninhabited island of Borokua,[5] in the archipelago referred to in Marovo as Vechala (the Russell Islands), and on the active volcano of Savo Island. The warriors from Marovo Lagoon maintained alliances with both Vechala and Savo. The local world-view of the time has a large-scale, inter-island scope (Hviding 2014a). In the mid-nineteenth-century heydays of headhunting, Tusu Marovo was iconically known for having four hundred warriors in residence (which would indicate a total population then of 2,000 or more) and was by far the strongest settlement of the lagoon and of the old district of Ulusaghe.

Nineteenth-century Tusu Marovo, then, was a formidable regional power, not only through the raids that its 400 warriors (with allies near and far) could accomplish. The island's inhabitants were true 'saltwater' people, possessing as they did very little land and practising only modest agriculture, but controlling the lagoon and seaways and building regional power through maritime mobility, raiding and trading. Labour was extracted as tribute from nearby groups of so-called 'bush people' and from captives taken back alive from overseas raids, and the island was known to produce extraordinary quantities of rings and ornate carvings from fossilised giant clam shell (*Tridacna*), the regional currency and ceremonial objects of old New Georgia. Such slavery-based large-scale production of valuables also took place among Tusu Marovo's close relatives and allies at the Marovo-speaking stronghold of Bili, by the southeastern entrance to the lagoon (Somerville 1897: 364) (see Figure 4.1).

To top it all, the Tusu Marovo people controlled (as they still do today) more than 50 km² of lagoon and a 10-km long section of the raised barrier reef, providing for abundant supplies of fish and other marine resources, and in the old days also providing for exclusive ocean entitlements to deep-sea fishing grounds for highly prized tuna. Tuna fishing, to which was added a reported massive breeding of domestic pigs on adjacent lagoon islets (not usual among the 'saltwater people' of the day), and the access through barter and tribute to large quantities of taro cultivated in irrigated pond fields by

---

5   There was no water on this island because Lagiti had previously stolen it from the mythical fruit bat Vekuveku.

the 'bush people' in the interior of Vangunu, enabled Tusu Marovo and its long-standing resident polity to arrange spectacular feasts to which guests were invited from throughout the New Georgia area (Bayliss-Smith and Hviding 2012). On the little island itself, the expansive groves of *Canarium* nuts and the fertile volcanic soil made for voluminous seasonal production of leaf packets of smoke-dried nuts, a regionally prized commodity then as it still is today.

In its heyday between about 1850 and 1885, Tusu Marovo had the Babata settlement of the Kalelupa group as its central stronghold, with the most powerful chief of the island resident there, several large war canoe houses doubling as halls for feasts and rituals, and the island's most important ceremonial sites nearby. Arranged along the first ridge above the coast, in an easterly direction towards the Marovo peak, were several smaller settlements, each managed by a leading warrior answering to the Babata chief. At the island's northwestern corner, the small peninsula of Olovotu was the location of a smaller stronghold managed by a sub-group of the Kalelupa, with a war canoe house and a small but strong fortified enclosure built on a rocky islet on the nearshore reef flat.[6] At the opposite side of the island, the eastern lagoon-facing coast had a series of settlements and infrastructures of the Kalekogu group, whose people also lived on some lagoon islands directly to the east. It was the Kalelupa and Kalekogu settlements together that could muster the proverbial 'four hundred warriors with club and shield', and that maintained at least four war canoes between them, with the added possibility of hiring more canoes from allies in the lagoon, since 400 warriors could crew much more than Tusu Marovo's own maximum of three or four war canoes (Hviding 2014b: 112). The spatial configuration of the settlements, and the provision of a continuous lookout from the summit, provided for commanding views of the lagoon from the deep passage of Jae to the southwest, through which canoes from the western parts of the New Georgia islands would have to enter, over to the northern ranges of the lagoon, across all major entrances through the barrier reef into the central lagoon, and far to the southeast towards Bili. On a smaller geographical scale, the separation of Tusu Marovo from the mainland by the comparatively shallow Gaio Passage made for easy monitoring against incursions from that side. It was said that from its peak Tusu Marovo had

---

6   See Somerville (1897: 390–2) for a description of this 'sacred islet'.

'eyes that could see everywhere', and no enemies could approach unseen, whether from the eastern Roviana side, northern New Georgia, Choiseul or Isabel, the southeast lagoon or even from the landward side.

Nowhere else in the eastern parts of New Georgia was there such a force, which was comparable only to the Nusa Roviana stronghold (Walter and Sheppard 2000). The scale of the Tusu Marovo polity reached beyond the island not only through the regular, successful raids of its warriors, but also through the alliance with satellite communities of relatives that had developed, since the early nineteenth century, at the deep barrier reef passage of Bili in the southern lagoon and at the barrier island of Ramata in the northern lagoon (see Figure 4.1). Both developed as enclaves of Marovo-speaking 'saltwater people' in areas dominated by 'bush people'. The Tusu Marovo–Bili–Ramata axis has remained significant into the present time through marriage, the seasonal convergence of all groups on Tusu Marovo for the harvest of *Canarium* nuts and the reaffirmation of kinship, and massive gatherings from near and far for funerals, as we shall see.

Signifying History: Experiencing Tusu Marovo

Oral history and an abundance of remains of stone terraces, fortifications and shrines combine to paint a portrait of what Tusu Marovo would have looked like around 1850. In the once central coastal settlement of Kalelupa at Babata, more than 15 houses were situated on platforms of coral rock in the tidal zone (see Figure 4.6), with ceremonial grounds and elaborate houses for the war canoes on the flat coastal land, shaded by trees. The Babata settlement, where the chief lived, commanded a sweeping view of Marovo Lagoon to the west, north and east. Further up in the lower hills were a significant number of terraced house sites constituting separate hamlets spread out in an east–west direction along the Kalelupa flank of the island. Higher up were fortifications into which women and children (and some warriors for 'security') would be directed from the Babata seashore settlement and the hillside hamlets when there was a danger of attack from enemies, primarily from the western districts of Roviana and Vella Lavella.

Figure 4.6: House on rock foundation in the tidal zone of Babata, Tusu Marovo, early 1890s

Note: Observe the small 'skull house' (or household ancestral shrine), fishing gear and canoe.

Source: Photograph by Lieutenant H.B.T. Somerville, 1894, reproduced with permission of the Royal Anthropological Institute.

From extensive hydrographic and ethnographic surveys over several months between 1893 and 1895, Royal Navy Lieutenant H.B.T. Somerville reported in some detail on the Marovo Lagoon, its places and its people. Under Admiralty orders, the 52-metre sail and steam ship HMS *Penguin* spent ten months of those years around New Georgia, while officers were often stationed for weeks and months in 'tenting parties' ashore near local settlements to survey inshore lagoon waters and reef passages. Somerville accordingly explains how he and his party of a dozen sailors was 'encamped' for some eight months altogether in different parts of the Marovo Lagoon. For Tusu Marovo, he noted how this 'hilly island only slightly detached from the coast in the eastern lagoon … was in old times the most populous and agreeable to trade at of any of the places nearby' (Somerville 1897: 360). Perhaps echoing the earlier navigational information given by Findlay (1877), Somerville (1897: 360) explains how '[f]rom this early communication [the island] has given its name … to all New Georgia on the older charts'. The maps from the survey of the New Georgia group produced during the HMS *Penguin*'s survey are detailed, and for Tusu Marovo they accurately show the settlements at the time along the northern coast, including, from east to west, Babata, Chochopo, Kinapoda, Mavara and Olovotu, as well as settlements on smaller adjacent lagoon islands.

Much later, from a visit in 1973, district officer James Tedder (1974) published a brief account of 'old village sites on Marovo Island' that gives an indication of the degree to which the Tusu Marovo landscape has remained riddled with indicators of how past generations lived on the island. To elaborate on this, my own notes from a walk of the steep northeastern parts of the island in the company of an elder and story-teller from Chea village (as well as some children, who came along for the excitement) provide the following account of the density and complexity of place on Tusu Marovo in pre-colonial times, as well as of how those layers of meaning persist today.

> As we climb up to Ta Lukutu, above where gardens are, there is a flat area with plenty of black basalt boulders, the soil mixed with stones. This is where a year ago, Risley found the extraordinary carved stone figurine which is reckoned to be associated with the story of Vekuveku. This is one of the few flat expanses of ground on the way up to Toa Marovo, and it was a place of worship in the old days. This is indicated by several old valves of giant clams lying around—they were used as temporary covers for ancestral skulls and to contain ceremonial artefacts. One of the small boys clears away some low shrubs and finds an extraordinarily large giant clam of a metre's length; we lift it, but there is nothing underneath it now, although a skull would have been there at some time. From Ta Lukutu, there is a clear direct view down to the small island of Ghireghire just off the shore, where the bones of ancestral generations are safekept, now covered by dense vegetation. Also, we have a sweeping view of the lagoon, right up to the Kalelupa people's western marine boundary in the Matenana Passage.

> We climb further up to Toa Kia, the place of *buli te Lagiti* (Lagiti's throwing sticks). There are four of these long, finely shaped basalt columns here, the largest one is seen on the steep slope below Toa Kia, broken into several pieces. At Toa Kia are many house foundations (*poi vanua*), and on the small expanses of level ground are a number of raised monoliths and *Cordyline* plants indicating ceremonial sites. The red and green *Cordyline* plants have been kept growing by the people until today to mark out the site as *hope* (prohibited, sacred).

We note one large flat stone slab of the type referred to as *langono*; several deep depressions are made into it for use in the manufacture of rings from fossilised clamshell.

As we climb on up the steep path there are many more house foundations, indicating dense settlement up here below Toa Marovo in the old times. These were the settlements of known ancestors, not of the unknown people who made the stone carvings at Toa Kia. Around the finely built stone foundations, the ground is riddled with old shells—*riki, roja, hulumu, nakolo, ropi*—the concentrations amount to middens, indicating how fond they were of bringing up food from the sea to cook it here. We are now right below the peak of Toa Marovo. According to the history of the Kalelupa people, this is where their founding ancestor Tutikavo first settled when he moved to the island with his two wives and a small number of warriors. His son Kelo then settled somewhat further down, and then Kelo's son Ireke started developing what would become the 19th-century coastal stronghold of Tusu Marovo in the tidal zone at Babata. The chiefs and their associates still maintained hill settlements, though, and it was not until the latter half of the 19th century that Babata became the permanent place of residence, following internal conflict and some dispersion of the Kalelupa people.

Approaching Toa Kiki, we come to the place where almost one hundred warriors from Bughotu in Isabel were massacred, after having been enticed up to the fortified hill settlement with promises of food and a feast. They came on a peaceful mission, it is said, and were first welcomed accordingly—according to Tusu Marovo traditions, an eloquent example of the past warfare strategies of the place, which valued deception, treachery, and ambush. Toa Kiki itself is a rugged black volcanic core of just a few m² devoid of vegetation, but surrounded by treetops, and can only be reached over a precariously narrow ridge. From here the view is spectacular, over the entire central area of Marovo Lagoon, indeed most of old Ulusaghe—towards Jae and Nono in the southwest, the Kolo mountains of New Georgia island, and the 1100-metre high crater rims of Vangunu. The entire lagoon coast is visible as far east as the Bisuana headland. Any approaching canoe can be seen from here. Right below Toa Kiki there are some house foundations again, but they are very old—it is not known who actually lived up here before Tutikavo came and settled nearby.

The vegetation all along the steep path leading up from the lower hillside gardens has a special character: it is filled with cultivated trees. There is an abundance of *Canarium* nut trees (as over most of Tusu Marovo), and additional fruit trees, hardwood trees for construction and canoe-building, decorative orchids, and a plethora of medicinal trees and shrubs, everything mixed with fast-growing creepers and climbers to form a rather dense forest of mainly small trees. This vegetation looks, feels, and smells very differently from the far less disturbed forest of the interior lands of the main islands of New Georgia and Vangunu. Apart from the characteristic tall *Canarium* trees with their white bark, which have been cultivated in groves by many generations of the island's inhabitants, the only really large trees here are a number of old banyans, whose tangled masses of trunks and branches hug the steep slopes and in places have invaded the stone structures of olden times.

Climbing a few precarious natural steps up to the Toa Gete we come out in the open, no trees obstructing the extraordinary view across ocean, lagoon, and land from this elevation of about 150 metres. When looking towards the distant barrier reef, the two villages of Chea and Chubikopi are in sight right down below and the terrain appears impossibly steep. During World War II, men of Tusu Marovo maintained a lookout here in cooperation with British coastwatchers to warn of any Japanese intrusion. On Toa Gete, as on several other peaks around the Marovo Lagoon, pyres would be lit as soon as Japanese vessels were sighted, and the message rapidly reached the coastwatcher base at Seghe. But elders say Toa Marovo looked somewhat differently then in the 1940s; there was a small expanse of flat ground on which a leaf shed was constructed to accommodate the lookouts—subsequent earthquakes have collapsed that flat area, so that Toa Gete is now surrounded by sheer cliffs and can only be climbed from one direction.

We leave the twin peaks of Toa Marovo and set off in a different direction towards the old coastal settlement at Babata. Over a series of old agricultural terraces (a precondition for cultivation on these steep slopes) we come to the dense stone structures of the fortified uphill area known as Bara Adoani ('enclosure to hide from view'). This is where women and children would be led when enemy warriors were approaching the island's shores. A rather extensive, quite flat area is contained here by finely built stone walls, some of which remain intact at about 1,5 metres' height. Bara Adoani had two entrances: a gate on the side for women and children, and a very narrow, fortified gate facing the path from below, through which only men are allowed to pass (still today, women do not pass through here, but follow another route down). Under siege, the central gate was guarded by warriors; enemies could only approach the gate one by one and would be killed as they did.

From Bara Adoani there are masses of old stonework in the form of house foundations and agricultural terraces, with a concentration in the area known as Raparapa (although today there are none of the leafy *rapa* trees here). This was a main settlement of the old times, also indicated by the proliferation of *Cordyline* plants which are markers of where people used to live. Here are many large, single *Canarium* nut trees, and very old secondary forest with a dense undergrowth of ferns and quite a few massive hardwood trees. The forest is interspersed with areas of more open vegetation of small trees and dense growth of tall gingers, indicating places of more recent cultivation. Although abandoned as a settlement long ago, Raparapa was in use much longer as a garden site. This is where women and children mostly lived in the old days. Until a permanent settlement developed at Babata, only men would venture down to the seashore, whether to fish, to travel on the lagoon, or to set out on raids in the war canoes that were kept in large elaborate canoe houses just behind the shore, hidden by the dense coastal vegetation. The establishment of Babata, probably from about 1850, changed that for there was no longer any need to hide from view given the sheer strength of the settlement.

At Vaributo, a small hill immediately behind the seashore, the Kalelupa ancestral shrine contains generations of chiefly skulls. We walk down past with hushed voices, reach the *Canarium* nut groves on the narrow coastal shelf, and through the coconut trees along the sandy beach we get a glimpse of the still imposing presence of old Babata. In the tidal zone close to the shore, about fifteen small islets of about equal size, now all overgrown by low mangroves, spread out along the beach for about 100 metres. They visualise a past when each was the stone foundation of a house, with small canoes, drying bark fibre nets and other fishing gears stowed around the buildings.

Piles of coral rock indicate the remains of causeways between house sites and beach. Immediately inland of the beach are remains of low stone walls indicating the location of the large ceremonial house in which war canoes were kept and rituals performed, including those involving skulls obtained through headhunting. On the beach facing two of the larger house foundations is the object most often shown to visiting tourists with the accompaniment of a chilling story: the *ngadoani veala* or 'stone slab for cutting the sacrificial child', where a child captured overseas would be killed and dismembered for subsequent cooking and ritual consumption.

Collapse and Re-Formations

At this stage we may return to the earliest European accounts, preceding the prolonged visits by the Royal Navy in the 1890s. In his navigational directory for the South Pacific, Findlay gives the following account, drawing on other accounts from early trading ships but echoing oral traditions about inter-island travel—and confirming prevailing pan-Marovo views about which local settlements were the most deeply involved in early stages of trade with Europeans, and about the nature of that trade:

> At [the west end of the barrier island of Matiu, in central Marovo] there is a narrow opening or passage where Sydney traders go in and anchor, to trade with the natives of *Repi* and *Marovo*, villages on the main island abreast this ... *The Marovo Passage* is about 4 miles off the mainland, and a number of small islets are dotted about among the reefs inside. Though this part of New Georgia is near 100 miles from Malayta, the natives of the two places communicate, making stopping places at Pavuhu [Pavuvu, in the Russell Islands] and Buaraqoi [Borokua]. There appeared to be but few inhabitants. Their canoes were stable, and strongly built. The natives were dark, sturdy looking men, and very active in the water, scrambling and diving in the most energetic manner. They brought off tortoiseshell and the usual ornaments, and were eager for pipes and tobacco.
>
> (Findlay 1877: 773, original emphases; with author annotations in square brackets)

Although Findlay did not quote his source for this information, it must be derived from the early traders ('Sydney' ones and others) who from about 1850 entered the lagoon through the reef entrance referred to by Findlay as the 'Marovo Passage', but known locally as the passage of Lumalihe— a shark-infested, turbulent deep channel with a maze of shallow reefs and sandy islands immediately inside it. Assuming Findlay's account is based on traders' accounts from about 1860 onwards, the village of Repi was just coming into existence after infighting had commenced within the stronghold on Tusu Marovo. Findlay's remark about 'few inhabitants' at the 'villages' of Repi and Marovo is puzzling, given that this is not far from, or perhaps still at, Tusu Marovo's period of climax when 'four hundred [men with] club and shield' could be mustered. It is, of course, possible that on the specific day Findlay's source paid a visit, the men of Tusu Marovo (and the 'renegades' at Repi) had other things to do than to paddle out to approaching ships; they were perhaps elsewhere on an expedition.

Wide-ranging transformations were soon to come. The large-scale political influence of Tusu Marovo was rapidly undermined near the turn of the century, when a combination of indigenous and imperial initiatives to end warfare, raiding and headhunting caused the rapid downfall of the regional economy and the political and ceremonial prominence of Tusu Marovo, as well as of other similar strongholds like Nusa Roviana. Prior to these interventions, a decline of the collective power of Tusu Marovo had set in from about 1885–1890, with a series of internal fights caused by accusations of incest and adultery as well as by mounting dissatisfaction with the Babata settlement's monopoly on contact with early European traders. From his survey work in the Marovo Lagoon in the 1890s, Somerville (1897: 399) reported that the 'once populous island of Marovo' had since 1885 been severely depopulated, but he probably overestimated the role of enemy attack in this process. In fact, several influential war leaders and sub-chiefs had been expelled from Tusu Marovo and had moved with their families and followers to coastal locations on the north coast of Vangunu. In the space of little more than ten years, powerful leaders who were used to owning war canoes and slaves, and managing warriors and overseas expeditions, found themselves without such power bases. Some retreated from view altogether to embark on a new life of shifting cultivation on the unlimited lands offered by in-laws from the district of Bareke, the north- and northeast-facing lagoon shores of the island of Vangunu, from which a number of chiefly women had married leaders at Tusu Marovo.

However, one new opportunity was offered to the downscaled former stronghold of Tusu Marovo: the missionaries. Methodist missionaries gained a foothold in southern Marovo in 1912, having already operated in the Roviana area for ten years. Tusu Marovo, although much weakened and with a precarious power base that included a dependence on good relations with traders and colonial officers, resisted this new presence until they were visited in 1915 by missionaries of the Seventh-Day Adventist (SDA) Church. Tatagu, the surviving chief and 'last headhunter' who had been forced to abandon the Babata settlement, offered the missionaries land near Olovotu, and that is how the mission village of Sasaghana was founded. With surprising speed, a generation of young men from kin groups throughout the central lagoon—groups that had until not long ago been prominent chiefly lineages and young men who were left with little to do as their old world had collapsed—enrolled in the SDA mission school established on Tusu Marovo. They were there to learn not only the Bible, but also (and this was deemed important by the old people) to learn English—in an indigenous quest for more equal footing with imperial power. Most of the first cohorts of Melanesian pastors of the SDA mission in the Solomons were from Tusu Marovo, and a new power base had been established for the island. Forty years later, after the Second World War, it was these once young men of Tusu Marovo, most of whom were born around 1900, who travelled to Australia and translated the Bible into Marovo (BSSP 1953). Meanwhile, as if the once-powerful ability of the small island to be efficacious (*mana*) was reconstituted through relations of control over the lagoon, its shores and its islands, a copra-based cash economy was expanded rapidly as new seashores within the realm of Tusu Marovo influence were cleared of forest and mangroves and densely planted with coconuts.

The early success of the Tusu Marovo people in so rapidly harnessing—in fact largely monopolising—the political-economic potential of the education-minded SDA mission led to a disproportionate number of people from the island—at first men, but gradually women as well—ending up with higher education. For a generation this was mainly pastoral training, but from the 1970s onwards a Tusu Marovo elite has been built up from both the Seventh-Day Adventists of Chea and Sasaghana villages on the Kalelupa coast and the Methodists from the Kalekogu village of Chubikopi. This educated elite has come to be mainly resident in the capital of Honiara and the Western Province capital of Gizo, occupied in education, government and business. From headhunting days to post-colonial government and modern-day business, the smallness of the island has never prevented its people from attaining influential positions.

Into the Twenty-First Century

With considerable shrewdness, Tusu Marovo's kin connections to other islands have been revitalised, in some cases even invoking kinship through captives taken on overseas raids many generations ago. In 1987 I witnessed a ceremony in which a man from the Bilua district of Vella Lavella was given a large piece of clamshell to signify his 're-entry' into the *butubutu* in recognition of two young women of his matrilineal group who had been abducted and brought as slaves to Tusu Marovo around 1880. Not unexpectedly, this reactivation of kinship links with Vella Lavella proved valuable because it involved access to large land tracts with timber resources from which logging royalties started to flow. Ultimately, in 2007, it was these logging-derived funds from the Vella Lavella connection that enabled the Kalelupa people of Chea village to build and open the largest church building anywhere in the Marovo region.

Figure 4.7: Part of Chea village, Tusu Marovo, 2012
Source: Photograph by Edvard Hviding.

Meanwhile, the various urban connections and the associated flow of money have seen an increase of rather lavish private houses being built back home, both for holidays and as retirement projects. Most named extended-family hamlets on Tusu Marovo, a collection of which constitutes each village, now commonly includes a diversity of leaf houses and permanent timber-and-iron buildings, as well as a seashore landing and a canoe house sheltering one or more fibreglass boats with large outboard motors (see Figure 4.7). However prestigious the hamlet's buildings and other property may be, the inhabitants are still likely to be engaged on an everyday basis in the subsistence economy of gardening and fishing, perhaps with a small-scale trading store or petrol depot on the side.

Socially speaking, Tusu Marovo does indeed remain the hub described by Wilson Liligeto (2006: 1) as 'the focal point of the entire Marovo Lagoon region'. At every funeral on the island, large numbers of people come from all around the Western Solomons to reaffirm their genealogical connections to the little island whose name still applies not only to itself but to the lagoon, to a political constituency, to an Austronesian language, and to a Melanesian people of about 14,000. In October 1991, the pioneer SDA pastor Kuloburu Liligeto (Wilson's father) died in Chea village at the age of 81. Although any funeral in present-day Marovo takes place only a day (or at most two days) after death, Pastor Liligeto's funeral saw more than 700 people coming to attend. Their courses by sea, or for some by air from the national or provincial capital, converged on the small island of large life. At the funeral, Chea elder and recognised genealogy expert Billy Kioto spoke to the gathering, invoking some powerful social imagery in characteristically high-level, old-style oratory to an audience with Marovo as the primary or secondary language:

> All of you who are here today and who speak the Marovo language—
> you are all from this place, and we are all kindred. Tusu Marovo
> here is where Marovo language comes from, and all those who speak
> Marovo today have come from this island, whether you are from
> Gatokae, from Viru, or from Gerasi, or from further away!

Today Tusu Marovo has the three large villages of Chubikopi, Chea and Sasaghana with a combined population of about 1,500 people, as well as a large diaspora in north and south Marovo, in Honiara and in Gizo. Belonging to two church denominations, the people of the island run two large primary schools. Since the 1990s tourist-oriented guesthouses have emerged in all three villages, but since the lagoon area has a rather broad range of resorts, tourism has not really caught on at the village level, and many inhabitants appreciate this fact, given that tourists are often seen as disturbing everyday life. A range of new infrastructure has been developed for Tusu Marovo in the twenty-first century, such as a reliable water supply piped from the mainland to all villages, replacing the unreliable supplies from poor freshwater springs, and a mobile telephone mast on the island of Mahoro opposite Chubikopi village. In 2010, the lofty height of the Toa Gete saw the installation of a government-funded solar-powered satellite dish receiver and wi-fi repeater that have since provided internet access to the schools of the central lagoon area, with spillover to the village people at large. The more recent expansion of mobile telephone services to include 3G coverage from four base stations throughout Marovo Lagoon has made the ownership and usefulness of smartphones very widespread.

Back at its roots of history, Tusu Marovo remains environmentally noteworthy for allowing for long-term continuous cultivation in its widely renowned rich black soil, as well as for having a remarkably intact forest cover, even if this consists of almost entirely domesticated secondary growth greatly enriched through arboriculture over successive generations. As such, the island, apart from its uncultivated southern slopes facing the mainland, is an almost fully anthropogenic environment. In marine terms, its inhabitants retain territorial control over an area of lagoon and barrier reef nearly ten times the extent of their land. Throughout my years of living on the island, I have been continuously reassured by its inhabitants that they have few worries about the sustainability of their lifestyle. With inexhaustible, fertile soil, the largest and most concentrated stands of valuable *Canarium* nut trees anywhere in the Western Solomons, abundant marine resources and good financial connections to town, business and government, Tusu Marovo with its dense population of about 250 per km² of land (not including sea) is indeed emblematic of a wider Melanesian pattern whereby islands that appear small are only small in a topographical sense.

Nevertheless, the largeness of the island is in other respects continuously challenged by political contest and, ironically, by conflicts between its residents and its diaspora. While in 2002 the Tusu Marovo people were able to demonstrate ancient matrilineal kinship to win a court case that gave them control over the financial outcomes of a major logging operation on the island of Vella Lavella far to the west, sharp conflict arose more recently between a small group of urban elite who proposed to log the entire Kalelupa side of the island—*Canarium* nut trees and all—and the residents (as well as other urban elite folk) who saw this as the ultimate folly. As one village spokesman expressed it: 'How can they imagine finding even a single commercially desirable big log here on this little island where every single tree has once been planted by someone?' Ultimately, in 2012, the Kalelupa people—who have established an impressive range of committees with constitutions, and even their own non-governmental organisation for 'sustainable development'—made the unprecedented move of disempowering their chief, a person who had been brought up in town, who had initiated the ill-fated logging proposal, and who, as was emphatically stated, 'cannot even speak the Marovo language properly'. In due course, the Village Committee of the Kalelupa people made the decision to install a new chief recruited from the wider circle of potential hereditary leaders. As it turned out, the non-Marovo-speaking chief suffered from poor health and passed away in 2013, after which a new chief was identified and installed. The new chief had spent part of his adult life in town but had much stronger local grounding, was capable in the vernacular, and was considered to have a good understanding of local needs and ambitions. In a quite unprecedented move, a public decision was also announced on the identity of his future successor. And in case anyone should come forward with objections, this decision already dealt with the genealogical complexities involved in the identification of *tuti bangara* ('the descent line of chiefly potential') (Hviding 1996: 143–6).

Conclusion

Life on the island continues to be large-scale, and few occurrences and phenomena appear to be entirely local. Post-colonial Tusu Marovo is at least as large as its pre-colonial version, and the vulnerability or resilience of its village life and diverse economies are, as in the old days, predicated on and generated from circumstances and relationships far beyond the island itself. In a present and future where relationships between people and place may seem less prescribed than before, the realisation is that the processes of education and migration in view today are not historically unique, but rather reflect truly long-term patterns for Tusu Marovo, in which knowledge and social organisation have always been situated between the endogenous and the exogenous, between expansion and contraction. The relational reach of Tusu Marovo remains large and open-ended. The island's own historian Wilson Liligeto (2006: xxii), whose vantage point is that of the Babata group but whose perspective embraces Tusu Marovo as a whole, may indeed be right in his gentle suggestion that '[o]ther tribes or people can learn from our experiences'.

References

Bayliss-Smith, T., and E. Hviding, 2012. 'Irrigated Taro, Malaria and the Expansion of Chiefdoms: *Ruta* in New Georgia, Solomon Islands.' In M. Spriggs, D. Addison and P. Matthews (eds), *Irrigated Cultivation of* Colocasia esculenta *in the Indo-Pacific: Biological, Social and Historical Perspectives.* Osaka: National Museum of Ethnology (Senri Ethnological Studies 78).

BSSP (Bible Society in the South Pacific), 1953. *Ia Buka Hope: Ria Tinototove Koina Oro Haguruna.* Suva: BSSP.

Duke, N.C., J.W. Udy, S. Albert and others, 2007. *Conserving the Marine Biodiversity of Marovo Lagoon: Development of Environmental Management Initiatives That Will Conserve the Marine Biodiversity and Productivity of Marovo Lagoon, Solomon Islands.* Brisbane: University of Queensland.

Findlay, A.G., 1877. *A Directory for the Navigation of the South Pacific Ocean: with Descriptions of its Coasts, Islands, etc., from the Strait of Magalhaens to Panama, and those of New Zealand, Australia, etc.: Its Winds, Currents and Passages* (4th edition). London: Richard Holmes Laurie.

Hubbs, C.L., 1951. 'Record of the Shark *Carcharhinus longimanus*, Accompanied by *Naucrates* and *Remora*, from the East-Central Pacific.' *Pacific Science* 5: 78–81.

Hviding, E., 1995. *Vivinei Tuari pa Ulusaghe: Stories and Legends from Marovo, New Georgia, in Four New Georgian Languages and with English Translations* (recorded, translated and edited by Edvard Hviding, with assistance from V. Vaguni and others). Bergen: University of Bergen, Centre for Development Studies, in collaboration with Western Province Division of Culture.

——, 1996. *Guardians of Marovo Lagoon: Practice, Place, and Politics in Maritime Melanesia.* Honolulu: University of Hawai'i Press (Pacific Islands Monograph 14). doi.org/10.1515/9780824851248

——, 2003a. 'Contested Rainforests, NGOs, and Projects of Desire in Solomon Islands.' *International Social Science Journal* 55: 539–554. doi.org/10.1111/j.0020-8701.2003.05504003.x

——, 2003b. 'Disentangling the *Butubutu* of New Georgia: Cognatic Kinship in Thought and Action.' In I. Hoëm and S. Roalkvam (eds), *Oceanic Socialities and Cultural Forms: Ethnographies of Experience.* Oxford: Berghahn Books. doi.org/10.2307/j.ctv287sjkp.9

——, 2005. *Reef and Rainforest: An Environmental Encyclopedia of Marovo Lagoon, Solomon Islands / Kiladi oro vivineidi ria tingitonga pa idere oro pa goana pa Marovo.* Paris: UNESCO (Knowledges of Nature Series 1). Revised and reprinted by UNESCO in 2011.

——, 2006. 'Knowing and Managing Biodiversity in the Pacific Islands: Challenges of Environmentalism in the Marovo Lagoon.' *International Social Science Journal* 187: 69–85. doi.org/10.1111/j.1468-2451.2006.00602.x

——, 2011. 'Re-Placing the State in the Western Solomon Islands: The Political Rise of the Christian Fellowship Church.' In E. Hviding and K.M. Rio (eds), *Made in Oceania: Social Movements, Cultural Heritage and the State in the Pacific.* Wantage: Sean Kingston Publishing.

——, 2012. 'Compressed Globalization and Expanding Desires in Marovo Lagoon, Solomon Islands.' In S. Howell and A. Talle (eds), *Returns to the Field: Multitemporal Research and Contemporary Anthropology.* Bloomington: Indiana University Press.

——, 2014a. 'Across the New Georgia Group: A.M. Hocart's Fieldwork as Inter-Island Practice.' In E. Hviding and C. Berg (eds), *The Ethnographic Experiment: A.M. Hocart and W.H.R. Rivers in Island Melanesia, 1908.* Oxford and New York: Berghahn Books.

——, 2014b. 'War Canoes of the Western Solomons.' In B. Burt and L. Bolton (eds), *The Things We Value: Culture and History in Solomon Islands.* Canon Pyon: Sean Kingston Publishing.

——, 2015a. 'Big Money in the Rural: Wealth and Dispossession in Western Solomons Political Economy.' *Journal of Pacific History* 50: 473–485. doi.org/10.1080/00223344.2015.1101818

——, 2015b. 'The Western Solomons and the Sea: Maritime Cultural Heritage in Sociality, Province, and State.' In E. Hviding and G. White (eds), *Pacific Alternatives: Cultural Politics in Contemporary Oceania.* Canon Pyon: Sean Kingston Publishing.

Hviding, E. and T. Bayliss-Smith, 2000. *Islands of Rainforest: Agroforestry, Logging and Eco-Tourism in Solomon Islands.* Aldershot: Ashgate Publishing.

Johannes, R.E. and E. Hviding, 2000. 'Traditional Knowledge Possessed by the Fishers of Marovo Lagoon, Solomon Islands, Concerning Fish Aggregating Behavior.' *Traditional Marine Resource Management and Knowledge Bulletin*, 12: 22–29.

Liligeto, W.G., 2006. *Bhata: Our Land, Our Tribe, Our People.* Suva: University of the South Pacific, Institute of Pacific Studies.

Magnuson, J.T. and R.M. Gooding, 1971. 'Color Patterns of Pilotfish (*Naucrates ductor*) and Their Possible Significance.' *Copeia* 2: 314–316. doi.org/10.2307/1442834

Somerville, H.B.T., 1897. 'Ethnographical Notes in New Georgia, Solomon Islands.' *Journal of the Anthropological Institute of Great Britain and Ireland* 26: 357–413. doi.org/10.2307/2842009

Stoddart, D.R., 1965. 'Geomorphology of the Marovo Elevated Barrier Reef, New Georgia.' *Philosophical Transactions of the Royal Society of London* B255: 383–402. doi.org/10.1098/rstb.1969.0017

Tedder, J.L.O., 1974. 'Notes on Old Village Sites on Marovo Island—New Georgia.' *Journal of the Solomon Islands Museum Association* 2: 12–21.

Walter, R., and P. Sheppard, 2000. 'Nusa Roviana: The Archaeology of a Melanesian Chiefdom.' *Journal of Field Archaeology* 27: 295–318. doi.org/10.1179/jfa.2000.27.3.295

——, 2017. *Archaeology of the Solomon Islands.* Honolulu: University of Hawai'i Press.

5

Manam Lives in Limbo: Resilience and Adaptation in Papua New Guinea

Nancy Lutkehaus

Manam Island, Volcano Island

As we know from the dramatic eruption of an underwater volcano in Tonga that created a new island in December 2021, the Pacific Ocean is far from pacific. The ash from the eruption—raining down from a plume that shot a mile and a half into the air—not only wrought havoc on nearby islands and caused planes to reroute, hindering much-needed aid to the Tongan people; it also sent tsunami-like waves across the Pacific Ocean and beyond. Indeed, because of the ocean's relationship to the so-called Pacific Ring of Fire, the region is characterised not only by its numerous islands, large and small, but also as a site of continual physical change. This dynamism results from the collision of tectonic plates that causes frequent volcanic and seismic activity (Torrence 2003; Connell 2015). Pacific islands not only appear, disappear and recede, they also *grow* due to volcanic and seismic activity— as is happening now on the Big Island of Hawai'i with the Kilauea volcano's continuing eruptions. Indeed, it was Kilauea's actions that originally led Pacific Island scholar Epeli Hau'ofa to contemplate the Pacific Ocean as a 'sea of islands' (Hau'ofa 1993). Other examples of active volcanoes in the Pacific, besides Hawai'i and Tonga, are found in New Zealand, Fiji, Samoa, Vanuatu, the Solomon Islands and Papua New Guinea (PNG).

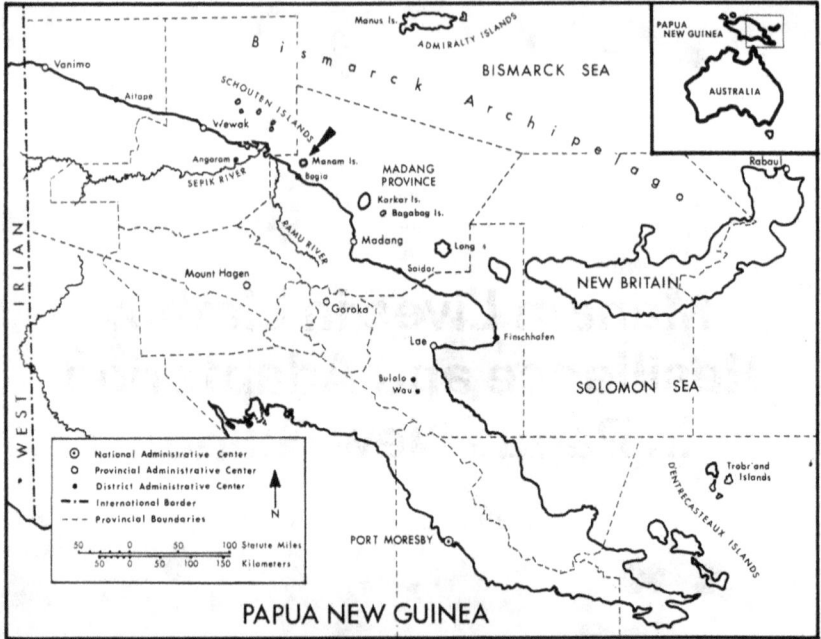

Figure 5.1: Papua New Guinea and Manam Island
Source: Lutkehaus 1995a: 2.

This chapter is concerned with Manam Island, one such small volcanic island in PNG.[1] A still active volcano located 13 kilometres off the north coast of Madang Province near the mouth of the mighty Sepik River (Figure 5.1), the island's lengthy and continuous eruptive history is recorded in the oral history of Sepik and coastal peoples (von Poser 2014), as well as in the earliest accounts of European explorers who traversed the north coast of New Guinea beginning in 1616 (Taylor 1958; Palfreyman and Cooke 1976; Johnson 2013). The number of currently active volcanoes in PNG is quite small, and among those the frequency of major eruptions is even smaller (Johnson 2013, 2020). Thus, as Filer points out in Chapter 2, active volcanic islands in the region are exceptions in a general discussion of small islands in peril.

We could conclude therefore that the disruptions caused by volcanic eruptions—like those that result from other natural disasters (tsunami, king tides, etc.)—are of less significance in the long term for small island

---

1   Regular updates on the condition of the Manam volcano are available from the Smithsonian Institution's Global Volcanism Program at volcano.si.edu/volcano.cfm?vn=251020

populations in general than other concerns such as population pressure. Nonetheless, their existence is still noteworthy, not simply because they create dramatic 'crisis' events that capture the media and the public's attention, but because, as I suggest here, there are lessons, both positive and negative, to be learned from them that have relevance to small islands in general. In the discussion that follows I focus on two contributions— one positive, the other negative—that an analysis of the fate of the Manam Islanders might have for other small island populations, as well as scholars and policy makers interested in thinking about the future of small islands in the Pacific. The positive contribution focusses on an ethnographic analysis of the successful grassroots adaptations Manam Islanders have made to a moment of extreme disruption to their lives. The negative contribution focusses on a broader analysis of the national and provincial plans to deal with this environmental disruption.

The 2005 Manam Eruption as a 'Moment of Extreme Disruption'

> [I]t has been an eventful week … Tavurvur resumed eruption on Monday evening and Manam produced another paroxysmal eruption on Thursday night … Our observation post at Warisi village was wiped out completely by what is described as pyroclastic flow … All our equipment was destroyed by the event. There were about 14 people at Warisi at the time of the eruption and all got injured while trying to escape …
>
> (Ima Itikarai, 29 January 2005, quoted in Johnson 2013: 311)

In January 2005, the entire population of Manam Island was dramatically evacuated as the result of a major eruption of the volcano that began in late 2004.[2] At the time there was extensive media attention that covered the event. Since their evacuation, the Manam Islanders have been living on the mainland across from Manam in what were meant to be temporary 'care centres' scattered along the north coast of Madang Province from Hansa Bay and Bogia, the provincial government's administrative centre, in the west to Mangem and Asuramba in the east (see Figure 5.2). Officially, the national and provincial governments have prohibited the islanders from returning to live on Manam as they claim that the extreme volatility of the volcano has made the island unsafe for habitation.

---

2   The reported size of the population that was evacuated from Manam ranges from 7,200 to 9,000 people.

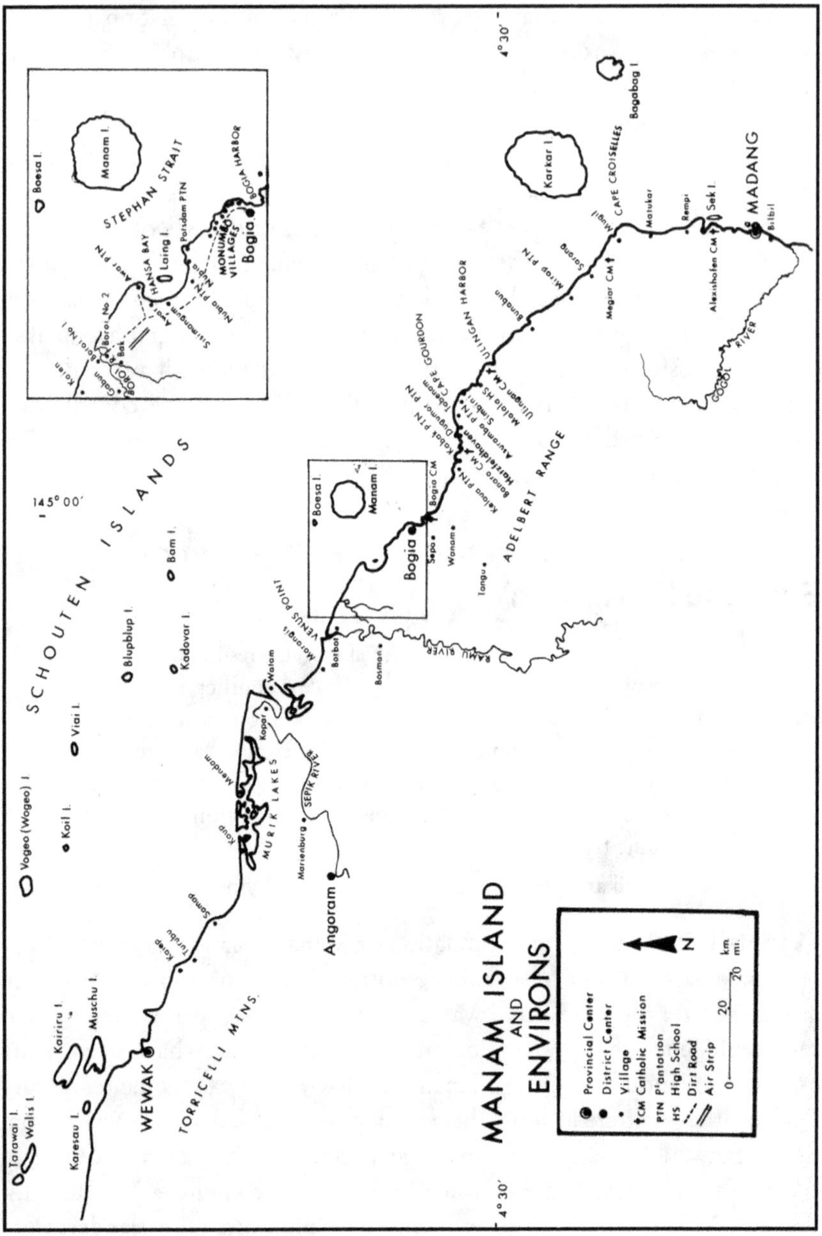

Figure 5.2: Manam Island and environs

Source: Lutkehaus 1995a: 38.

There is no doubt that the 2005 Manam eruption was an example of what Filer (Chapter 2) refers to as a 'moment of extreme disruption' that transformed the relationship between the 15 Manam Island communities and their island ecosystem in ways that were profound and irreversible. Although, as we shall see, some islanders have returned to Manam in defiance of the government's sanction (or, in one extreme case, at the behest of the provincial government), the population in the care centres, which has grown exponentially since 2005, remains 'in limbo'. They continue to live in the care centres while they await resettlement inland in Madang Province and their status continues to be that of what the United Nations (UN 2011) calls 'internally displaced persons' (IDPs).[3] This moment of extreme disruption in 2005 stands in contrast to previous moments of temporary disturbance caused by the Manam volcano that had occurred on the island before written colonial records began. A major contrast was that the PNG Government did not officially allow the islanders to return to live on Manam as they always had done in the past.

This means that the Manam Islanders have experienced a type of 'double jeopardy' in that they are subject to the same concerns that face the inhabitants of many small islands, such as population growth and resource depletion, along with the added concern of volcanic activity. Moreover, they have already been displaced from their island. How the Manam Islanders have adapted to their situation and why they are still in limbo is the focus of this chapter.

The Eruptive History of the Manam Volcano Before 2005

I have written elsewhere about the culture of the Manam Islanders prior to the major eruption in 2005 and their adaptation to living on an active volcano (Lutkehaus 1985, 1995b, 2016). In 1933 the British anthropologist Camilla Wedgwood also studied Manam society (Wedgwood 1933, 1934, 1937, 1948), as did the Catholic missionary Karl Böhm (Böhm 1975, 1983). Thus, we know quite a lot about the colonial and post-colonial culture and history of the Manam Islanders and the ways that they had adapted in the past to the periodic eruptions of the Manam volcano. Much like the

---

3   Vini Talai (2015) questions the validity or appropriateness of the use of this designation with regard to the Manam Islanders.

beliefs of the indigenous Hawaiians who attribute the volcanic activities of Kilauea to the Hawaiian goddess of fire Pelehonuamea, or Pele for short,[4] the Manam Islanders attribute the eruptive activity of the Manam volcano to a female spirit they call Zaria. They call the second, smaller crater Yabu, the name of a male spirit and Zaria's consort (Lutkehaus 1995a).

Figure 5.3: Nineteenth-century engraving of Manam Island
Source: Lutkehaus 1995a: 62. Reprinted from Finsch 1888.

The entire maritime region along the north coast of PNG in East Sepik Province and Madang Province is dotted with small islands, including the Schouten Island archipelago with the islands of Vogeo (Wogeo), Kadovar, Kairiru, Koil and Bam that are volcanic islands in the west and Karkar and Long Island to the east. Until the recent eruption of Kadovar in 2018, only Manam and the tiny island of Bam had experienced eruptive activity extensive enough to warrant evacuation of their entire island population in recent history. Bam last erupted violently in 1954 (Taylor 1960; Johnson

---

4   See Torgersen (2022) for a detailed discussion of the history of past and present indigenous Hawaiian beliefs about the goddess Pele and detailed references concerning Pele and her sisters. Pele is said to have originally travelled from Tahiti to the Hawaiian Islands. Beliefs in the relationship between supernatural spirits such as Pele and Zaria and volcanic activity are not unusual in the Indo-Pacific region. See the work of Michael Dove (2007, 2010) about the activity of Mt Merapi on Java for an insightful discussion of another example of the link between the moral dimensions of human–supernatural relations and volcanic eruptions.

2013). The Germans, who first colonised this region of New Guinea in 1884, named Manam *Vulkaninsel* (Volcano Island) in recognition of its defining characteristic (Finsch 1888). To the Manam Islanders, it is *Manam Motu* ('Manam Island') and the volcano is *Manam Ewa* ('Manam Fire').

The volcano rises majestically 1,807 m above sea level.[5] It is located in the southwest sector of the Pacific Ring of Fire, where the Pacific Tectonic Plate sinks beneath the Indo-Australian Plate. It is a textbook example of an almost perfect cone-shaped basaltic–andesitic strombolian volcano. In contrast to the usual more slow-moving effusive lava flows that erupt from shield volcanoes such as Kilauea, when a major eruption of a stratovolcano such as Manam occurs it causes a dramatic pyro-chemical explosion that shoots molten lava and rock thousands of metres into the air and rains down a thick layer of ash, blanketing villagers' gardens, trees and houses. Most of the time, however, there is simply a plume of smoke that hovers cloud-like over the mouth of the crater and trails across the sky (see Figure 5.4). The volcano's major vent forms its central crater with a smaller secondary vent to the north. Although these two summit craters are both active, most of the historical eruptions have originated from the southern crater, thus concentrating a major portion of the eruptive materials in the southwestern portion of the island. Four large radial valleys created by lava flows extend downward from the barren summit to the black sand beaches of the island. These 'avalanche' valleys are regularly spaced around the perimeter of the island and have channelled the flow of lava and pyroclastic material down to the coast (see Figure 5.5), creating dramatic 'frozen' lava formations that jut into the sea at the mouths of these valleys.

The island itself is 10 km in diameter. Before the 2005 evacuation of the island, its inhabitants lived in 15 villages situated along the black sand beaches of the island's periphery.[6] There were also two mission stations, Tabele and Bieng, run by Society of the Divine Word missionaries, which housed elementary schools run by the PNG Government on their premises. There was also a small health centre at the Bieng station that was operated by Catholic sisters (Lutkehaus 1995a, 1995b).

---

5   For further information, see volcano.si.edu/volcano.cfm?vn=251020

6   A smaller island named Boisa is located approximately 5 km off the northwest coast of Manam. It is home to 600 islanders who live in two villages. The Boisa Islanders speak their own dialect of the Manam language. They were not affected by the 2005 eruption and have remained on Boisa.

Figure 5.4: Manam Volcano, 1987
Source: Photograph by Nancy Lutkehaus.

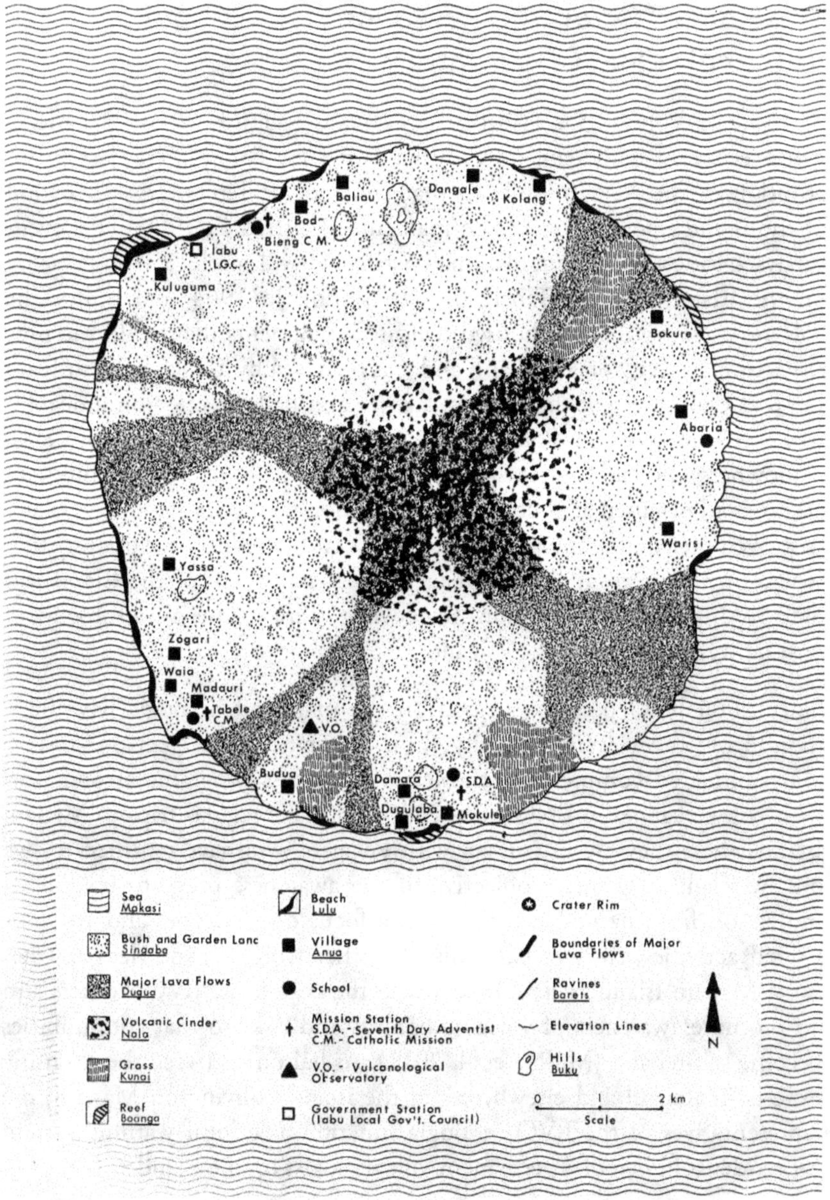

Figure 5.5: Manam Island villages and radial valleys formed by lava flows
Source: Lutkehaus 1995a: 45.

The Role of Scientific Evidence: The Rabaul Vulcanological Observatory

The Rabaul Volcano Observatory (RVO) was established in 1937 after a major eruption in Rabaul on the island of New Britain. The last major eruption of the Manam volcano that entailed the temporary evacuation of the island's population to the mainland occurred in 1959. This event catalysed the establishment of a small permanent vulcanological observatory on the island near the Tabele mission station in 1964 to monitor the volcano's activity daily. It housed delicate instruments that could monitor the seismic activity on the island and was operated by a vulcanologist from the RVO who lived at the Manam station and had direct communication with vulcanologists in Rabaul. However, a land dispute with the local landowners who had been leasing the land to the RVO led to the temporary closure and eventual abandonment of the monitoring station after vandals destroyed the equipment and the building that housed it. A second, smaller replacement station at Warisi village on the other side of the island was destroyed in the 2004 eruption, since when there has been no subsequent on-island monitoring of the volcano.

Since 2004 RVO personnel based on the mainland in Bogia have been monitoring the volcano's activity from a distance. RVO personnel have also conducted comprehensive volcano awareness programs at and near high-risk volcanoes in PNG since 2000 and make regular trips to Manam Island where they are in contact with villagers who have returned to live on the island. The main objective of the awareness program is 'disaster mitigation from natural hazards, with a focus on volcanoes and volcano-related activities, such as tsunamis'. In other words, acknowledging that some Manam Islanders have returned to the island, the RVO workers aim 'to promote awareness of volcanic hazards and risks so that communities become self-reliant' (Mulina et al. 2011). As liaisons between the scientific experts in Rabaul and elsewhere and the islanders living on Manam, the representatives of the RVO in Bogia function as a local warning system for residents in lieu of a permanent vulcanologist on the island.[7]

---

7   The Darwin Vulcanological Observatory and the Darwin Volcanic Ash Advisory Centre in northern Australia also closely monitor the Manam volcano, as does the Smithsonian Institution's Global Volcanism Program, because smoke emissions from the volcano can be hazardous to planes flying in the area.

Shortly after the major eruptions of the Manam volcano subsided, the vulcanologists at the RVO declared that it was safe for the population to return to the island. Thereafter they quietly and mysteriously withdrew their original statement about the safety of the islanders returning to Manam. The PNG Government then stated that it would no longer operate the two elementary schools on the island. The mission stations were also shut down and the Catholic priests and their assistants, the Servants of the Holy Spirit nuns, also left the island. The final blow to the once well-functioning island socio-economic system was the government's cessation of the copra boat that had operated on a regular basis shuttling people and cash crops, mostly copra and cacao, back and forth from the island to the mainland at Bogia.

The Establishment of the Manam Care Centres

Aid from fellow Papua New Guineans, the PNG Government, other governments worldwide, especially Australia and New Zealand, and international agencies such as the United Nations was provided to the displaced Manam Islanders immediately after the volcano's eruption and the evacuation of the island population to the mainland. This humanitarian assistance was especially prominent in the case of the Manam Islanders since the volcano's eruption came so soon after the calamitous disaster of the tsunami that hit large portions of Southeast Asia in December 2004.

The displaced islanders were housed in what were to be temporary 'care centres'—the UN's terminology for such emergency shelters that provide housing (usually tents) and aid—on the mainland at Potsdam, Mangem and Asuramba (see Figure 5.2). The care centres were established on land that had been transformed into coconut plantations during the colonial period, beginning with the Germans in 1884 and continuing with the Australians after 1914. After independence in 1975, the land had reverted to the Madang Provincial Government and was uninhabited at the time of the Manam eruption. Anxious to ensure that the Manam Islanders would not settle permanently on these lands, the original landowners, along with potential developers, have sought to reclaim them for the past 17 years.[8]

---

8   Connell and Lutkehaus (2016, 2017a, 2017b) have previously written about conditions in the care centres.

Short-Term Crisis Versus Long-Term Stasis

A series of stamps issued by the PNG Government depicts some of the largest and best-known volcanoes in the country. Mount Tavurvur, whose eruption in 1994 led to the destruction of Rabaul and the subsequent relocation of the East New Britain provincial capital to nearby Kokopo, is perhaps the most famous of these volcanoes.[9] Like the other stamps in the series, the Manam stamps present iconic images of an active volcano billowing clouds of dark but seemingly harmless smoke, exactly the kind of volcano—and dramatic image—that attracts volcano watchers, tourists, photographers and stamp collectors.[10]

The stamps featuring the Manam volcano were released in 2009, four years after the evacuation of the island's population. (Had the island been completely destroyed by the eruption it would probably not have been depicted on the stamps.) Significantly, the image on the stamps includes no people nor, of course, any reference to what has happened to the islanders. Instead, the stamps represent the romance and flirtation with danger that an active, but relatively quiescent, volcano offers outsiders who are far removed from the immediacy of the eruption and its aftermath. Thus, they do not reflect the real impact that the volcano's 2005 eruption, and more importantly, the government's inept response to it, has had on the lives of the displaced Manam Islanders.

The disconnect between the dramatic, but ultimately pastoral, images of the Manam volcano and the reality of the contemporary situation of the Manam Islanders also reflects another type of disconnection: the temporal distance between a moment of crisis and its aftermath. Once the immediate danger of the Manam eruption was over, and the immediate outpouring of humanitarian aid had been solicited, transmitted and dispersed, there

---

9   The resettlement of villagers living in the vicinity of Rabaul was used as a model for the potential resettlement of the Manam Islanders (Martin 2013).

10   There is an entire travel industry devoted to trips to volcanoes around the world. Like 'storm chasers', volcano tourists are drawn to the enticement of the dramatic and sometimes dangerous situations that active volcanoes and volcanic eruptions afford thrill seekers. See, for instance, the work of vulcanologist and photographer John Seach (volcanolive.com), which caters to film and television productions but also to 'adventure travellers'. Seach was one of the first outsiders to travel to Manam and take photographs of it erupting. In addition to Kilauea in Hawai'i, Yasur volcano on the island of Tanna in Vanuatu is the quintessential example of another active volcano in the region that remains an exciting but still relatively safe tourist attraction. The Yasur volcano was the focus of documentary filmmaker Werner Herzog's 2016 film, *Into the Inferno*, and appeared in the recent National Geographic documentary, *Fire of Love*. For many years adventure travel companies such as Mountain Travel Sobek have brought tourists to climb the Manam volcano as well.

has continued to be the much longer, less urgent, but in many ways more difficult period that has entailed the effort to return people's lives to some state of 'normalcy', whatever that new state might be. It is the difficulty of this decidedly unromantic and anti-heroic process, and some of the factors that have contributed to its complexity, that is highlighted here.[11]

And herein lies the crux of the cautionary tale that the Manam case provides to other small islands facing environmental pressures that are not tectonic in nature: the issue of resettlement. I will return to the complexities of resettlement later. At present I want to discuss the ways in which the Manam Islanders have dealt with their liminal status—that is, how they have breached the gap between the immediate outreach of aid they were provided at the moment of crisis or 'extreme disruption' in 2005 versus the problems caused by the failure of the post-colonial state to facilitate their resettlement 17 years later. What has filled that gap is the resilience, ingenuity and initiative of the Manam people themselves.[12]

Forms of Resilience and Adaptation

Over time, people inhabiting active volcanoes have developed various adaptations to living in such precarious environments.[13] The Manam Islanders had developed a social adaptation that served them well in the past when the volcano erupted and people needed to evacuate the island temporarily. Like many Melanesian peoples, they had a formalised system of exchange relationships. These exchange partners, whom the Manam call *tawa*, were individuals living in various villages on the mainland. This network of *tawa* relations was based on economic exchange and sociality that formed part of a system of regional integration that enmeshed islanders

---

11　As noted earlier, there was a huge outpouring of national and international humanitarian aid sent to the Manam people that helped them in the immediate weeks and months of their stay at the care centres. This benevolence is not to be dismissed lightly as it was much needed. But, as has been demonstrated time and again, the moral sentiments of humanitarianism—of heroic deeds being performed in the context of extreme natural disasters, which have themselves taken on increasing institutional and governmental importance globally—subside after time. Donor fatigue then leaves the more sobering reality of the complexities of dealing with the long-term effects of the initial events (Fassin 2011: 181–99).

12　See Oliver-Smith (1996: 312–4) for a similar argument in a cross-cultural context.

13　See Torgersen (2022) for a fascinating ethnography of how residents in the Puna region of Hawai'i's Big Island have adapted to living near Kilauea volcano in response to recent eruptions in 2014 and 2018. Unlike the Manam volcano, Kilauea is a shield volcano with long, low slopes, and thus poses slightly different threats to local residents. In both instances, people have chosen to remain living near the volcano and have developed adaptations, both mental and practical, that allow them to continue to live in a precarious environment.

and mainland people up and down the north coast of PNG (Lutkehaus 1985). For example, the Manam Islanders had exchange partners as far away as the Murik Lakes near the mouth of the Sepik River (Lipset 1997). In the past, *tawa* offered their Manam exchange partners temporary places of refuge and assistance with food and other material goods when they had to evacuate the island because of volcanic eruptions. Such relationships with exchange partners are not unique to the Manam or other people living on active volcanoes but are a key characteristic of many island populations.[14]

However, in 2005, due to the government's prohibition on the return of the islanders to Manam, the circumstances between them and their *tawa* changed. Given the extended period the islanders have remained on the mainland, these partnerships, which had been based on short-term stays in the past—a few weeks or months rather than years—predictably deteriorated. The traditional reciprocity that the Manam would have extended to their mainland *tawa*—in the form of pigs and baskets of galip nuts, as well as hospitality on Manam, was no longer possible. Not only were the islanders unable to reciprocate, but they had also overstayed their welcome.

After having lived in what were to have been temporary care centres on the mainland since January 2005, by 2011 the Manam Islanders had developed several grassroots adaptations to their precarious situation. As a result of the inability of the PNG Government to enact the promised resettlement, the islanders have both adapted to and resisted this condition by returning to the island. Given the proximity of the island to the mainland, many Manam Islanders return periodically to cultivate new gardens, to plant and harvest betelnut palms, raise pigs, harvest breadfruit and galip nuts, or to produce copra and gather wood for the construction of homes at the care centres. In some cases, they have also returned to live in their former villages on a permanent basis, just as they would have done in the past, depending, of course, on the activity of the volcano.[15]

---

14 Indeed, these island–mainland exchange relations are a subspecies of the more general Melanesian cultural phenomenon of exchange partners, of which much has been written by anthropologists over the years, especially with regard to the large ceremonial gift exchange events found in the highlands of New Guinea. Strathern (1971) provides a classic example.

15 As of 2011, two American missionaries and their families from Ethnos360 (formerly known as the New Tribes Mission) have also returned to live on the island, building permanent wood-framed homes there.

After violence erupted on the mainland between Manam Islanders from Baliau village and their mainland neighbours, and in order to reduce the potential for more conflict, a local provincial government official ordered the Baliau chief to move the entire village of 1,000 people back to their village on Manam (Aihi 2009; Matbob 2009). Powerless in the face of the conflict, the provincial officer told me that he had no choice but to remove the islanders from their temporary settlement on the mainland and the only option available was for them to return to Manam despite the national government's prohibition on doing so. Since the government still does not provide schools, transportation to and from the island, or health care for these villagers on Manam, the people of Baliau took it upon themselves to pay for schoolteachers for their children in the village. This example of islanders using their personal funds to provide what should be a public service to PNG citizens, is a striking example of empowerment and initiative in the face of government inaction on the part of Baliau villagers who want to get on with their lives and secure the future of their children (Mercer and Kelman 2010; Peter Irakau, personal communication, May 2015).

Thus, despite the lack of regular government-sponsored boat transportation between the mainland and Manam, there is a continual flow of people travelling back and forth to the island in small privately owned and operated banana boats with outboard motors. The Manam themselves have developed their own independent transportation network of these water-borne 'public motor vehicles' that carry people across the 15 km distance that separates those islanders living on the mainland beaches from their own island.

This system of transportation, now using outboard motors and fibreglass boats rather than traditional outrigger sailing canoes outfitted with coconut fibre sails (and later with outboard motors), is a modern technological transformation of the type of Manam-mainland connections via the sea that had always existed. It has become a small-scale local business opportunity for those individuals who can afford to purchase boats and outboard motors to service their Manam clientele.

Significantly, in the care centres, the Manam Islanders have maintained their traditional hierarchical political structure based on hereditary village chiefs called *tanepwa labalaba* ('big chief') (Lutkehaus 1990, 1995a). Some chiefs have requested that their relatives, especially their brothers, uncles and sons—men who have the title of *tanepwa sikisiki* ('small chief')—accompany villagers who want to return to live permanently on Manam.

Thus, they maintain claims to ancestral lands and resources on the island while the rest of their villagers remain in the care centres. Given that there has been an explosion in population growth in the care centres over the past 17 years (the population is said to have almost doubled), the return of villagers to the island has helped ease the cramped living conditions in the care centres. By sending members of the chiefly clan to live in their former Manam villages, the Manam chiefs are assuring the replication of the traditional Manam political structure in both the care centres and on the island. Thus, for example, the chief of Zogari village sent his uncle and his eldest son back to Manam to rebuild their homesteads in Zogari along with other families that have joined them to establish new gardens.[16] In other words, the chiefs have established two villages—one on the mainland and another back on Manam.

The Revitalisation of Traditional Exchange Festivals (*Buleka*)

When I visited Manam Islanders from Zogari village at the Potsdam care centre in 2015, they had just completed the construction of a new men's house (*keda*). This was a significant symbolic, as well as material, accomplishment since its completion was the catalyst for a traditional inter-village pig exchange festival (*buleka*) and recognition of a new village chief. Thus, the event signalled the revitalisation in a new location of the types of political, social and economic activities that traditionally took place on Manam. The preparations and the performance of the *buleka* served as a social activity that integrated various dimensions of Manam culture as a dynamic whole (Lutkehaus 1995a: 269–316). After some delays, the *buleka,* with its series of accompanying dances and distribution of pigs and food (taro, bananas, sweet potatoes and galip nuts), finally took place. Thus, at one of the care centres, Manam Islanders had begun to resurrect important political and economic activities, symbolic as well as constitutive of the social fabric of Manam culture, albeit in a new location and under less-than-desirable conditions.[17]

---

16 In addition to returning to live in former villages, some Manam Islanders have returned to inhabit the ruins of the buildings abandoned by the Catholic mission, including the former churches, and also the government schools, staking out claims as squatters on what had been traditional village land before colonisation.

17 I do not have information about whether or not similar *buleka* events have occurred at the care centres at Mangem and Asuramba.

A final impetus to Manam Islanders' returns to their island is their desire to bury their dead on Manam. Many islanders believe that it is important for their deceased relatives to be buried on the island, specifically near their family homestead. For these individuals, the role that *Manam Motu*—the land, the volcano and the sea that surrounds the island—plays in their spiritual and cultural life is so important to them that they have adjusted to the continuing periodic threats of volcanic eruptions. It is important not only for the dead but for their living descendants that the soul of the deceased reside on Manam Island (Lutkehaus 2016).

The Manam Resettlement Project

If in the past Manam Islanders would have returned to their island after the major eruptions of the volcano had subsided, why, given the volcano's lengthy period of relative quiescence, has the PNG state not formally declared the island habitable again? From one perspective, the government appears to want to ensure that the Manam Islanders have a safe and secure future. That is why it has passed the Manam Resettlement Authority Act (2016) and secured funding for the Manam Resettlement Authority that has developed a Manam Resettlement Project (MRP). However, 'security', as we know, is a powerful concept that works as much on potentiality, on the fear of possible events, as on their actual occurrence (Foucault 1977; Collier and Lakoff 2021). While proposing to resettle the Manam Islanders in a permanently safer location, the government has also kept them in limbo, living in declining conditions of health, sanitation and well-being in the care centres.

Why might this be the case?

One answer, among several, may be the usefulness to the PNG Government of the issue of the resettlement of the Manam Islanders as a source of international funding and potential development projects. A fundamental aspect of the MRP is the infrastructure—a road connecting the coast to the new inland settlement, which entails cutting down wide swathes of forest, building bridges, clearing land and so forth—that a project of this scale requires. Roads, of course, have long been a basic element of many infrastructure projects (Harvey and Knox 2015). In the case of the MRP, the contract to build the road was given to a logging company that was already harvesting timber from the surrounding area. The connection between infrastructure and small islands in peril lies in the fact that resettlement of any sort, urban or rural, involves issues of infrastructure and the allocation of funds for its construction—issues that are inherently political and susceptible to corruption. New theoretical insights by anthropologists into the concept of infrastructure have prompted us to think about the ways in which infrastructure can be considered in temporal terms. This move compels us to consider infrastructure not just as material things, such as roads, pipes and electrical wires, but also as 'intangible networks that facilitate connections and exchange' and as processes that can be analysed in terms of their lifespans, multiple temporalities and multiple actors (Larkin 2013; Anand et al. 2018). Thus, if 'we think of infrastructures as unfolding over many different moments with uneven temporalities', as Anand and others suggest, 'we begin to see that social and political dimensions are as important as the technical and logistical' (Anand et al. 2018: 17). The discussion of the temporal dimensions of infrastructure, coupled with insights into the importance of considering 'the promise of infrastructure'—that is, investigating people's hopes for the future embedded in infrastructure—offer useful avenues for a deeper understanding of the human dimensions of infrastructure and its implications for issues of migration and resettlement.

In their effort to focus on process rather than product, Anand and others eschew a focus on the 'finished' product of a planner's map. However, I suggest that the yet-to-be realised MRP, in the form of a document, is itself a type of bureaucratic infrastructure. As such, it too is an object of ethnographic interest and can be examined as part of the larger infrastructural process that entails various political actors. I build here on the idea that infrastructure can not only be thought of in temporal terms, as a process, but also as an assemblage of ideas and people. As theorists of urban infrastructure have pointed out (Graham and Marvin 2001: 8), infrastructures are socio-technical assemblages that represent long-term accumulations of finance, technology, expertise, and organisational and geopolitical power. Moreover, I suggest, the capital expenses and embedded socio-technics that infrastructures represent—that is, the dynamic and congealed processes of organising finance, knowledge and power—apply as much to rural as to urban infrastructures.

The Manam Resettlement Project as Document and Infrastructure

The origin of the MRP was the issue of security—the need to produce a solution to a situation the government had deemed untenable: the return of the displaced Manam Islanders to safely live permanently on their island. Instead, the national government has promised the islanders that they will resettle them inland and facilitate the creation of a new Manam settlement.

Whether the eruption of 2005, and subsequent smaller eruptions since then, have warranted the government's prohibition on return to the island is difficult to determine. For example, there was a recent eruption of the volcano in April 2022 that showered ash on several village homesteads and gardens but did not cause major damage or lead to the evacuation of villagers to the mainland. The indeterminacy of the situation, and the debate it engenders, is exactly the point. It has allowed the government to refuse to resume services while providing it with the need—as well as the opportunity—to develop a project to permanently relocate the Manam Islanders.

A focus on infrastructure and the politics of security as loci of congealed political power and aspirations, rather than just material forms, brings other aspects of the MRP into view. As 'cost estimates' from a PowerPoint presentation of the MRP revealed, these elements include the opportunity to accumulate finance, procure technology and establish new organisational structures to oversee the enactment of the plan.[18] In 2016, the Manam Resettlement Authority Act allocated an initial sum of 6 million kina (then worth approximately 2 million US dollars) to the MRP. The overarching benefit to the national government is its access to what Naomi Klein (2007) has called 'disaster capital'—funds that have become available because of the need to respond to various natural disasters and the accompanying neoliberal political and economic changes such capital facilitates.[19] A secondary goal is aspirational—the plan offers the state an opportunity to incorporate the Manam Islanders into its larger national goals of economic development and societal transformation.

---

18   I was present at a slide presentation of the MRP made by Dr Boga Figa at the Madang provincial government offices in May 2015 to a group associated with the International Organization for Migration. Dr Figa is a member of the committee in charge of the implementation of the plan.

19   See also Collier and Lakoff's recent (2021) discussion of the rise of the emergency state in the mid-twentieth century as a new type of social formation tasked with issues of national security.

I have suggested that, in its material form, as a written document and visual PowerPoint presentation, the MRP is a type of infrastructure in and of itself. If we think of the plan—and its performative role in presentations made by government bureaucrats—as a stage in the life history of the entire project, it is a form of infrastructure whose function is the manifestation of a concept, a verbal and visual representation whose goal is to connect ideas with a particular reality. It represents an imagined future settlement. It is thus the first—and necessary—stage in a lengthy process that entails the creation of other forms of infrastructure, such as roads, bridges and deforested land for commercial farming. It is indeed, as Fennell has said of infrastructure, 'a thing that facilitates relations between other things ... a thing that facilitates other projects' (Fennell 2015). This outcome is not in and of itself necessarily negative or positive. However, the cautionary tale regarding the MRP and small islands in peril is the nature of the plan's proposed transformation of the way of life of the Manam people, and the encumbrances involved in the plan's realisation.

The MRP's proposal to create a new Manam community for 20,000 people (a projected 5,000 families) includes several classic infrastructural components: in addition to the construction of a 23-kilometre road and bridges that will connect the Andarum hinterland, where land has been obtained for the Manam, to the coastal administrative centre of Bogia (Figure 5.2), it also entails the creation of a new provincial government bureaucracy—the Ramu Development Corporation—to oversee the project, new buildings to house these government employees, and the clearing of 5,000 to 10,000 hectares of virgin jungle in order to provide arable land to lease to the Manam Islanders for subsistence and cash crop cultivation. Eventually, new schools and health care centres will be built to provide for the education, safety and well-being of the Manam population. As a slide devoted to the program's approach stated, 'economic development is adopted in this relocation program as a responsible approach'. Thus, the project also includes plans for the cultivation of smallholder cocoa plots and 'growth centre development', which was described as the creation of a new oil palm plantation where Manam Islanders could find employment. We might sum up the presentation's statement that '[s]ocio-economic viability runs parallel to importance of health and safety for IDP families to ensure livelihood is sustainable' by saying that, taken as a whole, the MRP presents a vision of the transformation of formerly autonomous island-based subsistence gardeners and fishermen living on traditionally held, communally owned lands, who augmented their household economy with the production of cash crops such as copra and cocoa, into land-bound, commodity-based, workers dependent upon wage labour, cash crop production and leased land. In other words, it is at best a typical Western-based model of economic development and modernity, and at worst an invitation to return to participation in an environmentally (and hence economically) questionable form of plantation economy (Tsing 2005; Tammisto 2016; West 2016; Bolgar 2018; Mitman 2019; Chao 2022).

'The Promise of Infrastructure': Waiting for the Future to Unfold

Even if the MRP is never fully realised or enacted as described in the plan, it still represents a particular modernist vision for the future of the Manam people and their place in PNG. In the case of the MRP, there is an ironic logic that entwines the former colonial-era coconut plantations in New Guinea, specifically those along the north coast of Madang Province that were first established in the German colonial period and later operated by Australian plantation owners, where the Manam are now living in care centres, with new forms of transnational agribusiness—especially the oil palm plantations being developed in PNG and West Papua.[20] Even if there never are oil palm plantations developed at Andarum, as the MRP proposes, the document still proposes this idea as part of the solution to the relocation of the Manam Islanders.

Given that the MRP was initially proposed in 2011, then revised in 2013, and still awaits enactment, the project has been subject to what Gupta (2018: 11) has called 'infrastructural suspension'. In a similar manner to his description of large infrastructural projects in India that are stuck in their implementation due to legal challenges, lack of funding, and corruption, the MRP has also been suspended due to the of lack of adequate funding and internal politics at the national and provincial levels—specifically, allegations of mismanagement of the project's funds (Anon 2019).[21] Moreover, the length of the interruption of the project—11 years since the plan was first developed, 17 years since the Manam were evacuated to the mainland care centres—speaks to both the marginality of the Manam

---

20 For examples of the controversies surrounding oil palm plantations and their detrimental effects on both the environment and the people living near and working on them in the Global South, especially New Guinea, see Tammisto (2016), West (2016) and Chao (2022).
21 The article reported allegations of fraud and misappropriation of the 6 million kina allocated by the PNG government to the Manam Resettlement Authority. To date, no one has been held accountable for the misappropriation of the funds.

Islanders and the relatively low economic returns envisioned for the state (see Lipset 2013). However, the MRP still represents a vision of and for the future. In this respect its function is aspirational. As Anand and others have suggested: 'rather than thinking of infrastructures as distanced or aloof hardware networks', we also need to consider how infrastructures shape and are shaped by 'human experiences and sentiments of hope, inclusion, violence, and abandonment' (Anand et al. 2018: 11).

In this respect, the MRP is a *promise* of infrastructure that offers the Manam a hopeful future. As such it represents both a gesture of aid as well as an acknowledgment of the government's responsibility for the welfare of the Manam Islanders by enabling them to have a viable future. However, the slowness of the process of building this new future is of paramount concern to the Manam Islanders, who continue to grow increasingly frustrated with the delays they are enduring as the national government seeks to oversee the funding for the MRP. At this point, the only progress that has been made is the construction of housing for the new government employees who will manage the plan (and the clearing of trees necessary for the construction of the road to Andarum). The irony of this situation is not lost on people living in bush-material houses in the care centres while 'modern' Western-style houses with running water, electricity and metal roofs are constructed for a new contingent of bureaucrats.

From Economic Autonomy to (Potential) Plantation Labourers

To fully understand the impact of the MRP at the local level— an infrastructural project based on the premise of providing security to an island population in peril—and its larger implications for small islands facing the prospect of migration and resettlement in the future, it is important to consider it in its temporal as well as its spatial and material dimensions. Specifically, we need to include the initial displacement of the Manam Islanders as well as their extended settlement in what were to be temporary care centres as part of the infrastructural process inherent in the long-term resettlement plan. For the funding to be secured and road building to advance, the livelihood and well-being of the islanders has been seriously impacted. They are now refugees in their own country, living in cramped quarters while waiting for a future promise of a new settlement to be developed, a plan that offers less economic autonomy than their past way of life on Manam Island. However, by 2015 people had become resigned to the fact that they were not going to be able to return permanently to Manam and that the best option was some variation on what was offered in the government's plan. It is an offer, if it does eventuate, that many older Manam Islanders have gradually and reluctantly come to embrace, seeing it as an opportunity for there to be 'two Manams', one on the island and one inland.

Conclusion: A Cautionary Tale in Perilous Times

In 2015, some Manam Islanders thought of this option in terms of continuing to have villages on Manam but also eventually having inland villages at the new Andarum settlement. In 2023, the reality is a different variation on the idea of 'two Manams'. Rather than the islanders retaining their rights to their land on Manam as well as gaining access to stable ground inland, what has developed instead is a different form of 'two Manams': one made up of a portion of the original inhabitants who have returned to their traditional villages on Manam, the other the temporary village communities, with the same names and indigenous political structure, in the care centres. There are no exact figures available as to the number of Manam Islanders who have returned to the island. The exception to this model of two Manams are the villagers from Baliau and Dugulaba who, due to violence on the mainland, lost their access to temporary settlements there and had to return *en masse* to Manam.

The 'limbo' this chapter refers to is the physical situation the Manam Islanders have found themselves in since they were prohibited by the PNG Government from returning permanently to their former way of life on Manam Island. Thus, they continue to live betwixt and between the unofficially sanctioned villages on Manam and the promised inland resettlement community. This situation has led to their liminal status with reference to their identity as 'internally displaced persons', or 'squatters' and 'interlopers' on other people's land.[22] And finally, there is the existential

---

22 Unfortunately, as the Manam situation of living in the care centres has dragged on, their displacement in these centres has become yet another example of what the political philosopher Agamben has identified with reference to refugee camps as an increasingly characteristic form of the 'political space of modern life'. He sees camps as a new and stable spatial arrangement 'born out of a state of exception that became a norm' (Agamben 1998: 119–23). Thus, the Manam care centres become another example of 'caution' with regard to the complexities of resettlement.

limbo of being suspended in a situation where they have little or no control over the implementation of the MRP as it is under the jurisdiction and leadership of the Manam Resettlement Authority.

The narrative described here of the mismanagement of the Manam Resettlement Authority funds, and the initial impetus on the part of the post-colonial state to use the situation of the Manam Islanders as an opportunity for the national government to request international funds for the resettlement plan, is certainly not unique to PNG, nor to small developing nations the world over, as was seen in the United States after Hurricane Katrina (Klein 2007).

Where the Manam case does relate to other small islands facing various perils, especially population pressure, is that the solution of migration and resettlement can be fraught with complexities, not the least among them the fact that such plans are inherently political (Oliver-Smith 1996; Lipset 2013; Connell 2015; Gharbaoui and Blocher 2018; Collier and Lakoff 2021). Given that the enactment of any *local* solution for the inhabitants of small islands will entail the involvement of actors at the provincial level, and often the national level as well, migration and resettlement become both infrastructural and political issues—which means issues of power and funds that can entangle the oftentimes powerless inhabitants of small islands in matters that are far more complicated, costly and time-consuming than what might initially have appeared to be the case. While the story of the failure, thus far, of the MRP offers a cautionary tale for the resettlement of other small island populations in the future, the case of the Manam Islanders also offers hope in terms of the resilience, ingenuity and initiative of small island populations in the face of natural and man-made adversity.

References

Agamben, G., 1998. *Homo Sacer: Sovereign Power and Bare Life*. Stanford: Stanford University Press.

Aihi, D., 2009. 'Madang Girl Beheaded.' *The National*, 22 June 2009.

Anand, N., A. Gupta and H. Appel, 2018. 'Introduction: Temporality, Politics, and the Promise of Infrastructure.' In N. Anand, A. Gupta and H. Appel (eds), *The Promise of Infrastructure*. Durham: Duke University Press. doi.org/10.1215/9781478002031

Anon, 2019. 'Manam Resettlement Funds Diverted.' *The National*, 28 June 2019.

Böhm, K., 1975. *Das Leben Einiger Inselvolker Neuguineas*. St. Augustin: Anthropos-Institut (Collectanea Instituti Anthropos Volume 6).

——, 1983. *The Life of Some Island People of New Guinea* (intro. and transl. N. Lutkehaus). St. Augustin: Anthropos-Institut (Collectanea Instituti Anthropos Volume 29).

Bolgar, C., 2018. 'Small Nation, Big Potential.' *Wall Street Journal*, 6 November 2018.

Chao, S., 2022. *In the Shadow of the Palms*. Durham: Duke University Press.

Collier, S.J. and A. Lakoff, 2021. *The Government of Emergency: Vital Systems, Expertise and the Politics of Security*. Princeton: Princeton University Press. doi.org/10.1515/9780691228884

Connell, J., 2015. 'Vulnerable Islands: Climate Change, Tectonic Change, and Changing Livelihoods in the Western Pacific.' *The Contemporary Pacific* 27: 1–36. doi.org/10.1353/cp.2015.0014

Connell, J. and N. Lutkehaus, 2016. 'Another Manam? The Forced Migration of the Population of Manam Island, Papua New Guinea due to Volcanic Displacement in 2005.' Unpublished report to the International Organization for Migration, Geneva.

——, 2017a. 'Escaping Zaria's Fire? The Volcano Resettlement Problem of Manam Island, Papua New Guinea.' *Asia Pacific Viewpoint* 58: 14–26. doi.org/10.1111/apv.12148

——, 2017b. 'Environmental Refugees? A Tale of Two Resettlement Projects in Coastal Papua New Guinea.' *Australian Geographer* 48: 79–95. doi.org/10.1080/00049182.2016.1267603

Dove, M.R., 2007. 'Volcanic Eruption as Metaphor of Social Integration: A Political Ecological Study of Mount Merapi, Central Java.' In J. Connell, and E. Waddell (eds), *Environment, Development and Change in Rural Asia-Pacific: Between Local and Global*. London: Routledge.

——, 2010. 'The Panoptic Gaze in a Non-Western Setting: Self-Surveillance on Merapi Volcano, Central Java.' *Religion* 40: 121–127. doi.org/10.1016/j.religion.2009.12.007

Fassin, D., 2011. *Humanitarian Reason: A Moral History of the Present*. Berkeley: University of California Press. doi.org/10.1525/9780520950481

Fennell, C., 2015. 'Emplacement. Theorizing the Contemporary.' Society for Cultural Anthropology website, 24 September 2015. culanth.org.

Finsch, O., 1888. *Samoafahrten: Reisen in Kaiser Wilhelmsland*. Leipzig: Ferdinand Hirt und Sohn.

Foucault, M., 1977. *Discipline and Punish: The Birth of the Prison*. New York: Random House.

Gharbaoui, D. and J. Blocher, 2018. 'Limits to Adapting to Climate Change Through Relocations in Papua New Guinea and Fiji.' In W.L. Filho and J. Nalau (eds), *Limits to Climate Change Adaptation*. New York: Springer International. doi.org/10.1007/978-3-319-64599-5_20

Graham, S. and S. Marvin, 2001. *Splintering Urbanism: Networked Infrastructures, Technological Mobilities and the Urban Condition*. London: Routledge. doi.org/10.4324/9780203452202

Gupta, A., 2018. 'The Future in Ruins: Thoughts on the Temporality of Infrastructure.' In N. Anand, A. Gupta and H. Appel (eds), *The Promise of Infrastructure*. Durham: Duke University Press. doi.org/10.2307/j.ctv12101q3.6

Harvey, P. and H. Knox, 2015. *Roads: An Anthropology of Infrastructure and Expertise*. Ithaca: Cornell University Press.

Hau'ofa, E., 1993. 'Our Sea of Islands.' In E. Hau'ofa, E. Waddell and V. Naidu (eds), *A New Oceania: Rediscovering Our Sea of Islands*. Suva: University of the South Pacific, School of Social and Economic Development.

Johnson, R.W., 2013. *Fire Mountains of the Islands: A History of Volcanic Eruptions and Disaster Management in Papua New Guinea and the Solomon Islands*. Canberra: ANU Press. doi.org/10.22459/FMI.12.2013

——, 2020. *Roars from the Mountain: Colonial Management of the 1951 Volcanic Disaster at Mount Lamington*. Canberra: ANU Press. doi.org/10.22459/RM.2020

Klein, N., 2007. *The Shock Doctrine: The Rise of Disaster Capitalism*. New York: Picador Press.

Larkin, B., 2013. 'The Politics and Poetics of Infrastructure.' *Annual Review of Anthropology* 42: 327–345. doi.org/10.1146/annurev-anthro-092412-155522

Lipset, D., 1997. *Mangrove Man: Dialogics of Culture in the Sepik Estuary*. Cambridge: Cambridge University Press. doi.org/10.1017/CBO9781139166867

——, 2013. 'The New State of Nature: Rising Sea-levels, Climate Justice, and Community-Based Adaptation in Papua New Guinea (2003–2011).' *Conservation & Society* 11: 144–158. doi.org/10.4103/0972-4923.115726

Lutkehaus, N., 1983. 'Introduction: Missionaries as Ethnographers.' In K. Böhm, *The Life of Some Island People of New Guinea* (transl. N. Lutkehaus). St. Augustin: Anthropos-Institut (Collectanea Instituti Anthropos Volume 29).

——, 1985. 'Pigs, Politics and Pleasure: Manam Perspectives on Trade and Regional Integration.' *Research in Economic Anthropology* 7: 123–144.

——, 1990. 'The *Tambaran* of the *Tanepwa*: Traditional and Modern Forms of Leadership on Manam Island.' In N. Lutkehaus, C. Kaufmann, W.E. Mitchell and others (eds), *Sepik Heritage: Tradition and Change in Papua New Guinea*. Durham: Carolina Academic Press.

——, 1995a. *Zaria's Fire: Engendered Moments in Manam Ethnography*. Durham: Carolina Academic Press.

——, 1995b. 'Missionary Maternalism: Gendered Images of the Holy Spirit Sisters in Colonial New Guinea.' In M. Huber and N. Lutkehaus (eds), *Gendered Missions: Women and Men in Missionary Discourse and Practice*. Ann Arbor: University of Michigan Press.

——, 2016. 'Finishing Kapui's Name: Birth, Death and the Reproduction of Manam Society.' In E. Silverman and D. Lipset (eds), *Mortuary Dialogues: Death Ritual and the Reproduction of Moral Community in Pacific Modernities*. New York: Berghahn Books. doi.org/10.2307/j.ctvpj7hc4.13

Martin, K., 2013. *The Death of the Big Men and the Rise of the Big Shots: Custom and Conflict in East New Britain*. New York: Berghahn Books.

Matbob, P., 2009. 'Where Do They Go from Here?' *Post-Courier*, 10 July 2009.

Mercer, J. and I. Kelman, 2010. 'Living Alongside a Volcano in Baliau, Papua New Guinea.' *Disaster Prevention and Management* 19: 412–422. doi.org/10.1108/09653561011070349

Mitman, G., 2019. 'Reflections on the Plantationocene: A Conversation with Donna Haraway and Anna Tsing.' *Edge Effects* Magazine, 18 June 2019. edgeeffects.net/wp-content/uploads/2019/06/PlantationoceneReflections_Haraway_Tsing.pdf

Mulina K., J. Sukua and H. Tibong, 2011. 'Report on Madang Volcanic Hazards Awareness Program, 6–29 September 2011.' Rabaul: Rabaul Vulcanological Observatory.

Oliver-Smith, A., 1996. 'Anthropological Research on Hazards and Disasters.' *Annual Review of Anthropology* 25: 303–328. doi.org/10.1146/annurev.anthro.25.1.303

Palfreyman, W.D. and R.J.S. Cooke, 1976. 'Eruptive History of Manam Volcano, Papua New Guinea.' In R.W. Johnson (ed.), *Volcanism in Australasia*. Amsterdam and New York: Elsevier Scientific Publishing.

Strathern, A., 1971. *The Rope of Moka: Big Men and Ceremonial Exchange in Mount Hagen, New Guinea*. Cambridge: Cambridge University Press. doi.org/10.1017/CBO9780511558160

Talai, V., 2015. 'Political Ecology of Hazards: A Case Study of Manam Internally Displaced Persons.' Unpublished paper for the Pacific Research Colloquium, State, Society and Governance in Melanesia Project, The Australian National University.

Tammisto, T., 2016. 'Enacting the Absent State: State-Formation on the Oil-Palm Frontier of Pomio (Papua New Guinea).' *Paideuma* 62: 51–68.

Taylor, G.A.M., 1958. 'The 1951 Eruption of Mount Lamington, Papua.' Canberra: Bureau of Mineral Resources (Bulletin 38).

——, 1960. 'An Experiment in Volcanic Prediction.' Canberra: Bureau of Mineral Resources (Record 1960/74).

Torgersen, E.H., 2022. Lavaland: Vernacular Seismology in Volatile Volcanic Environments in Puna, Hawaiʻi. Bergen: University of Bergen (PhD thesis).

Torrence, R., 2003. 'What Makes a Disaster? A Long-Term View of Volcanic Eruptions and Human Responses in Papua New Guinea.' In R. Torrence and J. Grattan (eds), *Natural Disasters and Culture Change*. London: Routledge.

Tsing, A.L., 2005. *Friction: An Ethnography of Global Connections*. Princeton: Princeton University Press. doi.org/10.1515/9781400830596

UN (United Nations), 2011. *Protecting the Human Rights of Internally Displaced Persons in Natural Disasters: Challenges in the Pacific*. Suva: Office of the UN High Commissioner for Human Rights.

von Poser, A.T., 2014. *The Accounts of Jong: A Discussion of Time, Space, and Person in Kayan, Papua New Guinea*. Heidelberg: Universitaetsverlag Winter (Heidelberg Studies in Pacific Anthropology 2).

Wedgwood, C., 1933. 'Girls' Puberty Rites in Manam Island, New Guinea.' *Oceania* 4: 132–155. doi.org/10.1002/j.1834-4461.1933.tb00097.x

——, 1934. 'Report on Research in Manam Island, Mandated Territory of New Guinea.' *Oceania* 4: 373–403. doi.org/10.1002/j.1834-4461.1934.tb00119.x

——, 1937 'Women in Manam.' *Oceania* 7: 401–428. doi.org/10.1002/j.1834-4461.1937.tb00395.x

——, 1948. 'Trade and Exchange of Goods on Manam Island.' *Man* 48(5): 8. doi.org/10.2307/2792028

West, P., 2016. *Dispossession and the Environment: Rhetoric and Inequality in Papua New Guinea*. New York: Columbia University Press. doi.org/10.7312/west17878

6

Pressures and Perils in the Stony Bits of Lihir, Papua New Guinea

Nicholas Bainton and Colin Filer

Introduction

Small island communities are often under pressure or even in a perpetual state of peril. Narratives of climate change have focussed greater attention on the plight of Pacific Island communities (Barnett and Campbell 2010; Connell 2015), but it is worth recalling that some of these communities were already thought to be facing perilous futures in the colonial period, well before anyone was concerned about the effects of global warming. Despite the growing focus on rising sea levels and environmental change, many small island communities face a range of other problems. It is therefore important to understand how people have adapted to these challenges in the past, and how their options have been changing over time, in order to understand how they might adapt to the effects of climate change and other pressures in the future. In broad terms, the vulnerability of small island communities is shaped by a combination of factors, including crude population density, the physical characteristics of individual islands, the stocks of natural resources or ecosystem services owned by the inhabitants of these islands, and their access to goods and services beyond their territorial shores.

The Lihir group of islands in Papua New Guinea (PNG) has attained a certain notoriety because of the development of a large-scale gold mine on the main island of the group. In the 1980s, prior to the development of the mine, the smaller islands in the group were already experiencing the pressures of population growth. Mining activities commenced in the mid-1990s, and since then all sorts of social, economic and environmental changes have taken place as a result (Bainton 2010). The smaller islands have also experienced new forms of pressure that set them apart from all the other small island communities of PNG, where there is no huge gold mine in the immediate vicinity. Mining activities are currently expected to continue until at least 2040, which leaves plenty of time for these pressures to become even more pronounced in the absence of any affirmative action.

In this chapter, we examine the pressures and perils confronting the smaller islands in the Lihir group. Our chapter forms the second half of a broader analysis of land relations and pressures in Lihir, with the first half focussed on the direct and indirect forms of displacement created by the gold mining operation, both on the main island and the smaller islands (Bainton et al. 2022). Here our primary aim is to examine the relationship between the main island and the small islands, and the way that this relationship shapes local responses to pressure on the ecosystems that sustain the local population. Drawing on our combined experience of researching the social impacts of the gold mine, we present a long-term picture of these insular relations and pressures.[1]

The Moment of Exclusion or Inclusion

When the first social impact study was initiated in 1985, PNG's Department of Minerals and Energy and the Kennecott-Niugini Mining Joint Venture were both inclined to regard the shores of the main island as the outer limit of the area in which the social impacts of a future mine would be experienced (Filer and Jackson 1986: 148).[2] That is essentially because the exploration licence granted to this mining company only covered the main island, while another mining company was granted a licence to explore the

---

1   One of us (Filer) was lead author of the original social impact assessments for the mining project, while the other (Bainton) first came to Lihir as a PhD student in 2003, when the mine was already operational. Both of us have continued to reflect on the social impacts of the mine, from one vantage point or another, since we paid our initial visits.

2   The first social impact study was undertaken at the request of the New Ireland Provincial Government.

mineral resources that might be contained on one of the smaller islands. But it soon became apparent that the islanders themselves were not prepared to accept the right of the national government to make such artificial jurisdictional decisions. Indeed, most of them did not acknowledge the authority of the national government in any shape or form. Instead, a majority were members of a body called the Nimamar Association, which might best be described as a micro-nationalist movement (May 1982). The name of this body was an abbreviation of the names of the four islands in the Lihir group—Niolam, Malie, Masahet and Mahur. The members of this micro-nation regarded the state of PNG as an alien and unwelcome form of intrusion on their own sovereignty. Indeed, they already regarded the future advent of a large-scale mine as the fulfilment of a prophecy made by their own founder (Filer and Jackson 1986: 242–9).

There is no reason to believe that Lihirians had any conception of themselves as a singular micro-nation before the advent of European colonial rule. Yet they do possess a single vernacular language that is not shared with any other indigenous communities in New Ireland Province (Neuhaus 2015). During the 12 years that elapsed between the initial discovery of gold deposits in Luise Caldera in 1983 and the start of mine construction in 1995, local community leaders elaborated a variety of ways in which to assert their own collective identity and their claims to ownership of the mine located at the heart of this 'imagined community' (Bainton 2010).

Outsiders have sometimes applied the name Lihir to the main island on which the mine is now located. Lihirians themselves think of it as the name of their language group. The name of the main island that figures as Niolam in the name of the Nimamar Association is more accurately rendered as Aniolam—a name that translates as 'Big Island' or 'Mainland' in the Lihirian language. The other three names that figure in the name of the Nimamar Association are the names of three small islands located at various distances to the northeast of Aniolam (see Figure 6.1). In the 5 km of sea that separate Malie from Aniolam there is one very small island (Sanambiet) and one miniscule island (Mando) that are linked to Malie itself at low tide. Malie, Masahet and Mahur are known collectively as Ihot—a name that translates as 'Stony Places' or 'Rocky Places' in the Lihirian language.

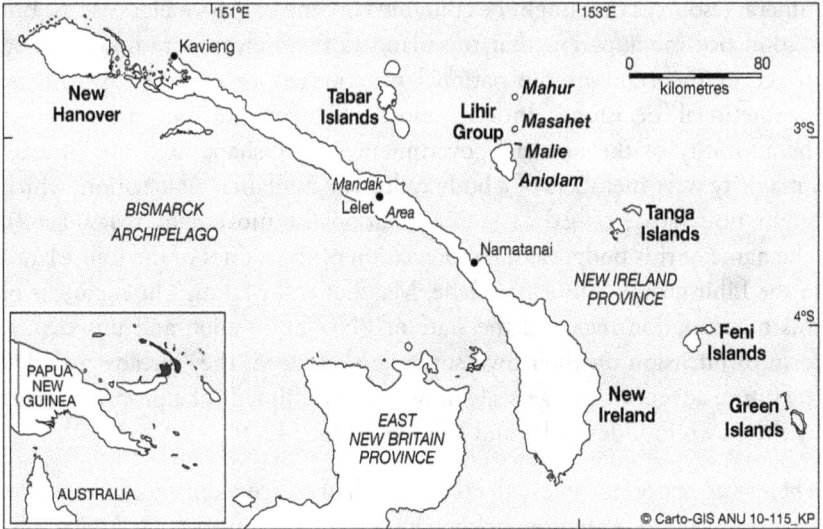

Figure 6.1: The Lihir group of islands
Source: CartoGIS Services, College of Asia and the Pacific, The Australian National University.

Small Islands Under Pressure

The island of Aniolam formed around five extinct volcanoes. It has a generally mountainous terrain, with dome-shaped peaks and razor-backed ridges separated by deep gullies, meaning that much of the available gardening land is located in coastal areas. By contrast, the islands of Malie, Masahet and Mahur are all raised coral shelves with flat tops, where crops are cultivated, and a narrow coastal strip, where most of the settlements are located (see Figure 6.2). A survey conducted in 1990 identified more than 400 separate hamlets across the whole island group, most of which were located within 200 m of the shoreline (Filer 1992a). Many of these hamlets are separated from each other and the surrounding vegetation by walls constructed out of slabs of coral limestone. This material is so abundant on the small islands that walls have also been built to separate clan territories and individual garden plots.

Figure 6.2: Masahet Island

Source: Photograph by Nicholas Bainton.

A calculation of the average number of rural villagers per square kilometre of land on each of the four islands in 1980 shows that densities were much higher on the small islands than on the main island (see Table 6.1).[3] A subsequent study found that malaria was 'meso-endemic' in settlements constructed on limestone soils, including all the small island settlements, but 'hyper-endemic' in settlements constructed on volcanic soils, including many of the settlements on the main island (Taufa et al. 1991). The obvious explanation for this difference is the lower rate of water retention on the limestone soils, but the variable incidence of malaria also helps to explain the higher population densities in this type of environment.

Table 6.1: Rural village population density in 1980

Island	Area (km²)	Population	Density
Aniolam	198.6	3,639	18.3
Malie	2.2	265	120.5
Masahet	8.1	841	103.8
Mahur	7.8	626	80.3
TOTAL	216.7	5,371	24.8

Source: Filer and Jackson 1989: 37.

---

3   The areas specified in this table have been slightly adjusted in light of more recent calculations based on satellite imagery. The landmass of Aniolam has also expanded slightly as a result of the dumping of mine waste rock into Luise Harbour during the course of mining operations.

Ninety per cent of the land on the small islands was already being used for the cultivation of food crops or cash crops before the construction of the mine. The rest was either occupied by human settlements or was otherwise unsuitable for cultivation. Part of the mountainous interior of the main island showed no sign of having been cultivated at any time in the past, so the average number of villagers per square kilometre of land 'in use' was more like 25 than 18.

A survey of village agricultural systems conducted in the mid-1990s, before the start of mine construction, drew a broad distinction between the system found on the small islands and the one found on the main island. The key difference between the two systems was the length of the fallow period in the swidden farming cycle. On the main island, a patch of land would be cultivated for 1 or 2 years and then left fallow for at least 5 years, and often for as long as 15 years, before being cultivated again. On the small islands, the fallow period was only 3–5 years in length, so the fallow vegetation did not have time to approximate anything that could be described as secondary forest. This difference in the intensity of cultivation, which clearly reflects the relative scarcity of cultivable land on the small islands, was accompanied by some differences in the relative significance of the food crops that were being cultivated. On the main island, the most important crops were the short yam (*Dioscorea esculenta*) and sweet potato, followed by the long yam (*Dioscorea alata*), cassava, banana and taro. On the small islands, sweet potato, cassava and both types of yam were of equal importance. In both systems, there was a difference between those gardens in which a planting of yams might be followed by a planting of other root crops, and those in which two plantings of sweet potato and/or cassava were made before the land was left fallow (Hide et al. 1996).

Further evidence of the relative significance of different food crops in these two agricultural systems is available from a national nutrition survey conducted in the early 1980s, since the Lihir group of islands was included in the survey sample (see Table 6.2). The greater preponderance of sweet potato and cassava in the diet of the small islanders, and the complete absence of taro, is further evidence of the growing scarcity of arable land, since both crops produce greater yields from soils of lower quality. These figures also indicate the relative importance of coconuts in the small island system, not only as a source of food but also because the very limited availability of fresh water meant that villagers were drinking a lot of coconut milk and were especially reliant on it during periods of drought or dry weather.

Table 6.2: Percentage of households consuming different food crops on previous day in 1982

System	Coconut	Sw. potato	Yam	Cassava	Banana	Taro
Main island	62	58	46	28	17	9
Small islands	81	94	58	39	6	0

Source: Hide et al. 1996: 66, 77.

The authors of the first social impact study for the future development of the mining project conducted their own survey of the main island's agricultural system at the end of 1985. They calculated that 19 food gardens on Aniolam ranged in size from 900 to 2,800 m², and that the main constraint on garden size was the need to build a bamboo fence around the perimeter in order to keep the pigs out. Since the fences began to deteriorate after nine months, the food crops had to be harvested within that period, leaving tree crops like breadfruit, banana and papaya to keep growing as the garden began to revert to fallow (Filer and Jackson 1986: 39). They also reckoned that the area of land in fallow at any one time was roughly 7 or 8 times the area under active cultivation, which led them to estimate that each person required a minimum of roughly 5,000 m² (or half a hectare) of cultivable land in order for the system to provide an adequate food supply for households that could not afford to purchase food from elsewhere (ibid.: 40).

When the social impact study was being revised and updated in 1988, the authors conducted an additional survey of eight food gardens on the island of Masahet, and found that these were smaller, on average, than those on Aniolam, ranging in size from 440 to 700 m². The higher density of the human population was found to be associated with a lower ratio of pigs to people and a more concerted effort to separate the pig habitat from the areas of settlement and active cultivation. On Masahet, a combination of stone walls and bamboo fences was built in order to divide the whole island into three concentric zones, and thus confine the pigs to the area of secondary forest between the coastal zone of human settlement and the interior plateau where short-fallow swidden cultivation predominated. Where stone walls had been built to protect gardening areas, tubers and vegetables could be cultivated for a longer period without the risk of pigs breaking through rotting fences (see Figure 6.3). Furthermore, since the fallow period was much shorter than on the main island, various measures had been adopted to compensate for the desiccation and degradation of the soil (Filer and Jackson 1989: 39–40).

Figure 6.3: Garden stone wall, Masahet Island
Source: Photograph by Simon Foale.

In 1992, the mining company commissioned another survey of village agriculture, with a sample comprising 16 households in three villages on Aniolam and 6 households in one village on Masahet (Bonnell 1992). This survey found that 86 people on Aniolam accounted for 38 gardens under cultivation, with a combined area of 5.436 hectares (632 m^2 per person), while 50 people on Masahet accounted for 50 gardens with a combined area of 2.405 hectares (481 m^2 per person). These figures suggest that the total area of land cleared for new swidden gardens in the decade preceding development of the mine would have been 250–300 hectares per annum (around 1.5 per cent of the total surface area) on the main island, and 85–100 hectares per annum (or 5–6 per cent of the total surface area) on the small islands.

Despite the differences in average garden size and the duration of the swidden cycle discovered by subsequent surveys, the authors of the first social impact study suggested that both agricultural systems required the same amount of cultivable land per head of population—about half a hectare—in order to be sustainable (Filer and Jackson 1986: 40). On this basis, they calculated that more than 60 per cent of the 'usable land area' would already have been required to feed the resident population of two of the small islands—

Malie and Masahet—in 1980, and if the population was growing at an average rate of 2 per cent per annum, the supply of cultivable land would have run out before 2020 (ibid.: 41).

This does not mean that the islanders would run out of food; it simply means that they would have to import more food from elsewhere, and so would have to find the means to pay for it. In this respect, the villagers on the main island, who still had plenty of cultivable land, also had an option that was ceasing to be available to those on the small islands. They could sell cash crops, or surplus food crops from their gardens, or some of the domestic pigs that made a nuisance of themselves, or else the products of their hunting and fishing practices, and use the proceeds to buy some of the other things they needed. The 1980 national census gave some evidence of the extent to which they were able to exercise this option (see Table 6.3).[4] In 1985, the average cash income from the sale of export crops (copra, cocoa and chillies) across the whole of Lihir was thought to be approximately 40 kina (then equivalent to 40 US dollars) per capita per annum, but the average would have been considerably lower on the small islands (Filer and Jackson 1986: 55).

Table 6.3: Sources of household income from own agricultural production in 1980

Island	Households	Copra (%)	Cocoa (%)	Chilli (%)	Veg. (%)	Fish (%)
Aniolam	906	41.9	43.3	0.8	18.9	14.7
Malie	71	5.6	0.0	21.1	4.2	2.8
Masahet	199	9.5	1.5	38.7	5.5	4.0
Mahur	139	53.2	3.6	28.1	2.9	1.4
TOTAL	1,315	36.3	30.4	10.5	14.4	11.0

Source: Filer and Jackson 1986: 51–2.

In 1968, it was reported that the Malie islanders were exporting 50 bags of copra each month from their coconut plantation on the tiny (50-hectare) islet of Sanambiet (Gormley 1968). Twenty years later, they were barely selling any copra at all. Some of the palms on Sanambiet had been cut down to make way for food gardens during a drought in 1982, despite the very poor quality of the soil, and coconuts were only being used for domestic consumption and pig fodder (Filer and Jackson 1989: 40). Even on Mahur

---

4   The figures for cocoa and chillies are for the number of households claiming to grow them, whether or not they sold them, while the other figures are for claims of sales.

Island, where more than half the households reported some copra sales in 1980, the market had dried up as the price of copra fell relative to the cost of hiring a motor boat to transport it to the nearest point of sale on the main island.[5]

The chillies being grown on the small islands were inter-planted with other garden crops, so did not have the effect of reducing the amount of land available for food production and did have the advantage of a fairly high value-to-weight ratio. The combined value of chilli exports from all three of the small islands was still around 30,000 kina in 1985, which would have yielded an average per capita income of around 15 kina, but the trade had entirely ceased by 1988 (Filer and Jackson 1989: 85). By that time, the most significant of the crops being traded out of Masahet Island were the pineapples that were also planted in regular food gardens (ibid.: 39). The harvest of marine resources such as crayfish and lobsters from the reefs around the small islands provided a regular supplement to the diet of the islanders, but there was no surplus left over for export, nor any local enthusiasm for using the small number of sea-going vessels as fishing vessels, given that they were mainly being used to transport people and cargo between the islands (Filer and Jackson 1986: 33). In 1985, one man described the production and export of shell money as the main 'business' activity on the small islands (ibid.: 93). In 1988, the annual value of this particular export was estimated to be 50,000 kina (Filer and Jackson 1989: 86). At that juncture, it seems safe to say that the islanders were making almost no money at all from any agricultural activity. Most of their cash incomes were more likely derived from the export of human beings.

The authors of the original social impact study could not find reliable figures on the rate of growth of the resident village population in the decades preceding the 1980 national census (Filer and Jackson 1986: 32–4). Subsequent discovery of a complete set of village books from 1958 has enabled us to compensate for this deficiency in the demographic record, and hence to compare the rates of change on the main island and the three small islands (see Table 6.4).

---

5   In any case, the only fibreglass motor boat based on Mahur Island had broken down and drifted all the way to the Solomon Islands in 1987 (Filer and Jackson 1989: 110).

Table 6.4: Rural village populations in 1958 and 1980

Island	1958			1980			% change		
	M	F	All	M	F	All	M	F	All
Aniolam	1,269	1,122	2,391	1,915	1,724	3,639	50.9	53.7	52.2
Malie	107	101	208	130	135	265	21.5	33.7	27.4
Masahet	314	270	584	439	402	841	39.8	48.9	44.0
Mahur	230	192	422	315	311	626	37.0	62.0	48.3
TOTAL	1,920	1,685	3,605	2,799	2,572	5,371	45.8	52.6	49.0

Source: 1958 village books and 1980 national census.

The first point to notice about these numbers is the excess of males over females on all four islands in 1958, especially on the small islands of Masahet and Mahur. The excess seems to have diminished in 1980, and even to have been reversed on the small island of Malie. However, these calculations do not take account of the number of individuals who were absent from the islands when the populations were counted. The 1958 village books only recorded three absentees, who are included in the figures shown here, although the true figure may have been higher, since the village books were meant to determine the amount of head tax to be collected from the residents of each village. The 'rural community registers' produced in advance of the 1980 national census were meant to include a more accurate count of the number of absentees. The New Ireland register showed that there were 510 individuals then absent from all the villages of Lihir—278 from Aniolam and 232 from Ihot. Since most of the absentees would have been males, the preponderance of males in the total village population was likely to have been just as great in 1980 as it was in 1958, which is indicative of high maternal mortality rates and low female life expectancies across all four islands.

Once absentee rates are taken into account, it would seem that the overall rate of population growth between 1958 and 1980 would have been much the same on the small islands as on the main island, although it still seems to have been lower on Malie, where the proportion of absentees was almost exactly the same as on Masahet—about 12.5 per cent of the total village population. The proportion of absentees was lower on Mahur—about 10.5 per cent of the village population—but still higher than on the main island of Aniolam, where it was only 7 per cent. This suggests that migration was already functioning as a kind of safety valve to relieve the pressure of a growing population on the limited natural resources of the small islands. We do not know enough about the absentees to know exactly how they came to leave the islands, but the authors of the social impact studies thought it was likely that many of the small islanders, in particular, had taken advantage of the educational opportunities offered by the Catholic Church in order to join the formal sector workforce in other parts of the country (Filer and Jackson 1989: 120–2).

The History of Island Symbiosis

There is clear evidence that the people of Ihot, especially those of Malie and Masahet islands, had been responding to the local scarcity of cultivable land and other subsistence resources for several decades—if not centuries—before the discovery of the gold deposit (Filer and Jackson 1989: 38–41). The intensification of their agricultural system was one of these responses. However, to understand the full range of responses, it is necessary to recognise the specific forms of symbiosis or mutual dependence that have characterised relationships between the small island communities and the larger population of the main island.

Father Karl Neuhaus established a Catholic mission station at Komat, on the southwestern coast of Aniolam, in 1931.[6] Shortly afterwards, he recounted the 'tragic' circumstances in which a man called Targolam from Masahet became the first Lihirian to be baptised (as Johannes) in 1907. Targolam was arrested by German government officials in 1906 because he was one of two men held responsible for leading an attack on a group of labour recruiters in 1905. A subsequent government report noted that the attack was most likely provoked by their 'recruitment' of a local woman against the wishes of her husband, which was said to be one example of a pattern of behaviour that was upsetting the local men because there was a 'pronounced lack of women on the islands'. Neuhaus concluded from his reading of the report that the attack was probably justified, but that Targolam was not the ringleader, and his arrest was mainly due to the fact that he and his fellow villagers were accused of attacking another village on Masahet in which 'they killed and ate various people'. It was the people of this village, one of whom had apparently been a member of the colonial police force, who encouraged the German officials to take revenge on their enemies. Targolam

---

6   He had first visited the islands in 1913 and continued to make regular visits after 1918.

was imprisoned at Namatanai, on the east coast of New Ireland, and died shortly after he was baptised. Neuhaus described him as the 'patriarch of Christianity' in Lihir (Neuhaus 1932).

This account is noteworthy for several reasons. The government report attributed the shortage of women on the small islands to a high rate of female mortality in a previous epidemic of dysentery, but since the shortage was still evident in 1958, it seems to have been a more enduring phenomenon, and would thus have limited the rate of population growth over a longer period. It is also somewhat surprising that different villages on an island as small as Masahet could function as autonomous political communities and engage in periodic acts of warfare, yet still survive to fight another day. Their mutual hostility might have been an effect of tensions induced by the advent of German colonial rule, but it is more likely that pre-colonial political communities in Lihir generally numbered no more than 200 people, and that Malie was the only one of the three small islands that was too small to contain more than one such community. Neuhaus did not say whether he believed that the people of Masahet were in the habit of eating each other, or how often they might have done so. However, Father Andrew Pong, who was the first Lihirian (and second Papua New Guinean) to be ordained as a Catholic priest, was quite fond of recollecting that his forbears from Malie would sail their canoes across to the main island, and then kill and eat various people when they got there. By the time the mining company arrived, Pong had established himself as the 'patriarch of Christianity' among the Catholic population of Lihir (Filer and Mandie-Filer 1998: 9).[7] And his emergence as a leader was not unrelated to the observation made by the early Catholic missionaries, that the people of the small islands were 'much more open and lively than the bush people from the big island' (Skalnik 1988, quoted in Filer and Jackson 1989: 172). Father Neuhaus might even have established his mission station on one of the small islands, if only there had been a harbour for his boat. In the 1980s, no visitor could fail to notice that the small islands boasted by far the most elaborate and finely decorated churches in the whole of Lihir, aside from the one at the central mission station (Filer and Jackson 1986: 108). And in 1985, the small islands accounted for 18 of the 31 Lihirians employed by the Catholic parish, including the teachers at the vocational school and the nurses at the health centre attached to the mission station (ibid.: 149).

---

7   Methodist missionaries had gained the adherence of a minority of the population on the main island in the 1930s (Hemer 2011).

Figure 6.4: Shells used to make shell money
Source: Bainton 2010: 96.

One might therefore be led to suppose that the relationship between the small islands and the main island was traditionally a relationship of domination or subordination, and that this relationship has simply acquired a novel Christian form. Yet there were some ways in which the people of Ihot have always relied on resources that could only be obtained from the main island and could not simply be obtained by force. For example, they do seem to have specialised in sailing long distances to neighbouring island groups as participants in a regional maritime trading network, but the timber required to make the canoes on which they sailed was most likely sourced from the main island. Before the Second World War, it was reported that some of them were sailing to the Tabar island group to exchange live pigs, tobacco and dances for shell money and food crops (Groves 1935: 360). These exchanges were part of a network through which shell money produced on the island of Lavongai (New Hanover) moved in a southeasterly direction, towards Bougainville, while the pigs generally moved in the opposite direction as the shell money was used to purchase them (George and Lewis 1985: 33; Bainton 2010: 94–9). But where did the pigs come from? In the Lihir case, the small island traders most likely used some of the shell money obtained from Tabar to buy pigs from people on the main island. In the 1980s, they were still using shell money (as well as cash) to buy pigs from this source, but after the Second World War they were also buying the shells as raw material from other parts of New Ireland and making the shell money themselves (see Figure 6.4).[8] And that is why they described the production of shell money as a kind of 'business', even if this business had also come to be seen as part of a wider enthusiasm for the maintenance of customary social institutions (Clay 1986: 193).

Nor were pigs and timber the only useful things that the small islanders got from the main island. In the 1980s, people from Malie and Masahet were obtaining bamboo poles and sago fronds from the northeastern part of the main island in order to construct the walls and roofs of their houses (Filer and Jackson 1986: 30, 42).[9] It is not clear how much they paid for these commodities, or whether they were able to obtain some of their supplies at no cost from close relatives. But when the mining company required bamboo for construction of their exploration camp, local villagers claimed that this was one of their 'cash crops' and proceeded to haggle over the price (Filer and Jackson 1989: 216).

At this juncture, some people from Malie and Masahet were certainly using relations of kinship and marriage to access additional gardening land in the northeastern part of the main island (Filer and Jackson 1986: 43). This was already a fairly well-established way of dealing with the shortage of land on the small islands. In the 1960s, a government official noted the social tension that arose when some people from Masahet went one step further and planted coconuts on the land they were gardening (Gormley 1968). The planting of tree crops is widely regarded as a way to claim use rights that extend for a lengthy period, but when those tree crops are also cash crops, the original owners of the land would be especially upset by the prospect of their loss. The general decline of copra production across the islands would have alleviated this particular source of tension by the 1980s.

The arrival of the mining company created a new opportunity for people on the small islands to enlarge the scope of what had long been their main way of alleviating pressure on their narrow resource base, which was for some of them to find paid work in other places and remit a portion of their incomes

---

8 The shells were purchased in bottles, then polished, drilled and strung together in spans or fathoms about 1.5 metres long. This was a group activity undertaken for several days at a time (Filer and Jackson 1986: 39).

9 The limited supply of bamboo on Malie was mainly used to stake yam vines, while the Masahet islanders were using a type of hibiscus, as well as imported bamboo poles, for this purpose. The people of Mahur still had sufficient local supplies of bamboo and sago for building purposes.

to relatives who stayed on the islands. As we have seen, some such work had been found on the main island, mostly through the good offices of the Catholic Church. The arrival of the mining company seems to have resulted in some diminution in this traditional source of employment. In 1985, the small islands accounted for 72 per cent of the wages that the parish paid to Lihirian workers, but by 1987 the figure had fallen to 53 per cent (Filer and Jackson 1989: 244). The fall may have been partly due to the fact that the exploration camp was a good deal closer to the small islands than the Catholic mission station.

The authors of the original social impact study estimated that 29 per cent of the people employed by the mining company for some period of time in 1985 came from the small islands, which was roughly the same as the proportion who came from Putput, the village adjacent to the exploration camp (Filer and Jackson 1986: 144). More significant, perhaps, was the fact that the small islands accounted for five out of the nine local 'staff' employed by the company at the end of 1985, since these were skilled workers who commanded higher wages and, unlike the unskilled casual workers, were entitled to accommodation in the exploration camp. This is significant because people from the small islands were initially excluded from the category of 'local' people who were meant to be given preference in the work of exploration (ibid.: 148).

Since unskilled casual workers were employed for relatively short periods of time and were not provided with accommodation by the mining company, they either had to accommodate themselves on nearby land to which they had some customary right, or else be hosted by local landowners living in the vicinity of the exploration camp. This was not an ideal situation for casual workers from the small islands. Twenty-one men from Masahet were briefly jailed in 1985 after two of them threatened to assault one of the expatriate geologists because the company was only providing them with one free meal at lunchtime each day, and they were not inclined to spend all their wages on the payment of rent to local landowners or the purchase of food from local trade stores (Filer and Jackson 1986: 126).

In 1987, the small islands still accounted for 20 per cent of the value of wages paid by the mining company to Lihirian workers, but the share going to the village adjacent to the exploration camp had risen to 46 per cent (Filer and Jackson 1989: 238). Most of the casual workers from the small islands were still 'squatting' on land that belonged to the local villagers (ibid.: 241), but their average wage was 38 per cent higher than that of workers from the main island (ibid.: 243). That was not because they were receiving a special accommodation allowance, but rather because they had higher levels of skill. In 1987, the estimated per capita income from employment on the project (by both the mining company and its contractors) was 40–100 kina on Masahet, 20–40 kina on Malie, and 10–20 kina on Mahur (ibid.: 240). It is hard to know how much of this money actually found its way back to the islands. However, when these sums are combined with the value of wages paid to islanders working in other jobs on the main island, or in other parts of PNG, it is likely that remittances from such sources were a far more significant source of cash income for households located on the small islands than were sales of shell money and pineapples.

Recommendations for Mitigation

The authors of the social impact study that was first drafted in 1986, and then revised in 1989, said that something had to be done to address what they described as 'an approaching ecological and demographic crisis' on the small islands (Filer and Jackson 1986: 26). This would be a key part of a wider program of action to mitigate the future social and economic impacts of the gold mine. Since the study was commissioned by a 'liaison committee' including representatives of the national government, the provincial government and the mining company, it was not clear who should take responsibility for which parts of this action program, but the general assumption was that the mining company would assume most of the responsibility once the development agreements were put in place.

The authors doubted whether the islanders could do much more to raise the productivity of the local agricultural system, with or without external assistance. They were likewise sceptical of the chances of persuading the islanders to adopt modern contraceptive practices, at least in the short term. That was not just because they were devout Catholics, but also because the church itself was the main provider of health services to its local flock. So they focussed their attention on the question of how the islanders might benefit from the prospect of further employment during the construction and operation of the mine. And that was because all parties were generally agreed, at least by the time that the impact study was revised in 1989, that indigenous Lihirians, including those born on the small islands, should have first preference in the allocation of jobs by the mining company and its contractors.

Given that the mine would be located on the main island, and that the customary owners of the mine lease areas were likely to get first priority in the allocation of all the 'benefit streams' that it generated, there was no question of recommending a form of affirmative action that would give any sort of priority to employment of Lihirians from the small islands, even if it could be argued that they were in greater need of it. So what was recommended was a set of measures to ensure that jobs would be fairly equally distributed between the different parts of Lihir, including the small islands and those parts of the main island that were furthest from the mine lease areas. The justification for doing so was that the customary owners of the area earmarked for mining operations would receive a share of the mineral royalties, as well as compensation payments for environmental damage, while people already living in fairly close proximity to the mine site would be in a better position to benefit from the company's obligation to procure goods and services from the locally based contractors who would be the targets of its business development program.

The basic issue here was the likelihood that the mining company would persist in its policy of only providing accommodation to local employees who were at a senior level in the corporate hierarchy. If workers at lower levels were expected to reside in their villages and commute to work on a daily basis, then those living in villages at some distance from the mine site would only be able to do so if the local transport network was substantially upgraded. Otherwise, they would either be marginalised in the local labour market or would make a nuisance of themselves by 'squatting' on other people's land, just as they had done during the early years of exploration (Filer and Jackson 1989: 347). And if the number of such people were to grow with the overall expansion of the mine-related workforce, so would the risk of conflict with the local landowners.

One of the earliest and strongest demands of community leaders across the main island was that construction of the mine should be made conditional on the construction of a 'ring road' around the whole island. If properly maintained, this would enable buses or trucks to convey workers from their villages to the mine site or the associated town site on a daily basis and take them home again after their shifts were finished. But what about the small islands? It was not hard to imagine the operation of a ferry service, although ferries would still have to travel a distance of 30–40 km to reach the most far-flung island of Mahur. The trouble was that none of the small islands had a jetty or wharf capable of accommodating such vessels, and even fibreglass banana boats had trouble navigating their way through the fringing reef, especially at low tide or in bad weather. It was in fact the government, not the mining company, which sought to deal with this problem by blasting passageways through the reefs in the 1980s. The authors of the social impact study wondered if the mining company could be asked to lend a hand in this enterprise in its capacity as a good corporate citizen (Filer and Jackson 1989: 340). But it was not clear what steps might later be taken to establish the infrastructure required for the operation of a proper ferry service or how that operation would be funded and managed.

Wholesale Economic Transformation

In 1985, the authors of the social impact study were presented with a letter from the leaders of the Nimamar Association, which they insisted was not a manifesto, even though it described their desire for a radical reordering of society that combined a form of millenarianism with specific economic and political objectives (Bainton 2010: 63–8). Nimamar members not only claimed that they had predicted the arrival of the mining company but also predicted that 'Lihir will become a city'.

And so it came to pass, in one way or another. Following a lengthy negotiation process, the mining agreements were eventually signed in 1995 (Filer 1995), and mine construction was completed by 1997. The development of the mine brought vast amounts of physical infrastructure and services to the island group. In addition to the special mining lease (SML) in Luise Caldera, three other leases were granted by the state in the northeastern corner of Aniolam to enable the development of the mining camps and townsite, and the airport that services the company and the residents of Lihir. An unsealed ring road was constructed to link the mining project and the town with all the coastal villages around Aniolam, fulfilling one of the demands from community leaders.[10] The development of the mine also required the relocation of households in the two coastal villages of Putput and Kapit to make way for the plant site and the stockpile within the SML. These were two of the five villages on the main island that were classified as 'affected area' communities due to their proximity to the project and their loss of customary land rights. Although the relocation of Kapit village has been an abject failure (Hemer 2016; Bainton et al. 2022), these newly defined affected area villages generally received the greatest share

---

10   It takes approximately three hours to drive all the way round the ring road.

of benefits, including improved housing, sealed roads, reticulated water supplies and electrical connections, and numerous mine-related payments. While these freshly developed villages may not have constituted a 'city', they certainly took the form of a model Melanesian suburb (Filer and Mandie-Filer 1998).

The development and expansion of the mining economy has brought wholesale economic change to Lihir. Through the payment of compensation, royalties, salaries and wages, and the awarding of local and national contracts, the mine has pumped several billion kina into the islands and the local region since operations first began. For example, between 1997 and 2018, the mining company paid out over 719 million kina in royalties, of which 20 per cent was paid directly to Lihirian landowners (Anon 2019). In 2019, the company paid out some 228 million kina in salaries and wages to more than 3,000 Papua New Guineans, many of whom were Lihirians. In 2016, the mining company reported that a total of 232.6 million kina had been paid to Lihirians between 2006 and 2015 under the terms of the primary benefit-sharing instrument known as the Integrated Benefits Package Agreement (Bainton and Macintyre 2021).

While these economic flows amount to a massive increase in per capita income, and a huge improvement in services and infrastructure, they have also irreversibly transformed Lihirian social relations as mine-derived wealth and mine-related impacts are unevenly distributed throughout the group of islands. Plain observation reveals that wealth is increasingly concentrated in the hands of a small group of male lease area landowners, most of whom reside in the 'mine-affected area'.[11] These inequalities have been the source of enduring and acrimonious conflict between, and within, the so-called 'affected' and 'non-affected' villages. It might therefore be argued that chronic economic inequality created by the mine represents a more perverse realisation of the dream to transform Lihir into a city. Some households, especially those in Putput village, are now wholly integrated into the cash economy, but most Lihirians residing on the outer islands, along with many households on the main island, still rely upon swidden agriculture to supplement incomes derived from the mining economy.

---

11   When the main compensation agreement was signed in 1995, there were 90 so-called 'block executives' recognised as the principal customary owners of these lease areas, and 75 per cent of them were living in the mine-affected villages on Aniolam (Filer and Mandie-Filer 1998: 4).

Small Islands Under Even More Pressure

Between 1996 and 2003, the social and economic impacts of mine construction and operation were monitored in a sequence of seven reports commissioned by the mining company (then Lihir Gold Ltd). The first four reports were compiled by Martha Macintyre; the last three by Martha Macintyre and Simon Foale. The consequences of population growth on the small islands of Ihot constituted one of the topics addressed in these reports.

The fourth report in the series included a graph showing the 'land needs' of the inhabitants of each of the small islands as compared with those of the main island (Macintyre 2000: 25). This graph was based on the earlier calculation that each individual would need at least half a hectare of cultivable land in order for the local agricultural system to provide an adequate food supply for households that could not afford to purchase food from elsewhere (Filer and Jackson 1986: 40). The graph showed that land needs on the island of Malie had already reached 130 per cent of the available land by 1999, while those on the island of Masahet were roughly 90 per cent. The graph was revised in each of the subsequent reports (Macintyre and Foale 2001: 14; 2002: 27; 2004a: 22). The last one showed that land needs on Malie were nearly 140 per cent of the available land by 2003, while those on Masahet were almost exactly 100 per cent, those on Mahur were more than 60 per cent, and those on Aniolam were almost 40 per cent. The most pessimistic predictions of the first social impact study were therefore confirmed (Filer and Jackson 1986: 41).

The graphs containing the calculations of 'land needs' may even have underestimated the extent of the problem by assuming that the area of cultivable land on each of the small islands was actually constant. Aside from the expansion of village settlements, there was evidence that garden fallows were subject to encroachment by rhizomatic weed species, especially 'kunai' grass (*Imperata cylindrica*) and the 'malo' fern (*Dicranopteris linearis*) (Macintyre and Foale 2002: 42). The spread of the fern was especially noticeable on Masahet, but also in some parts of what had now been designated as the 'mine-affected area' on the main island (ibid.: 149). The labour required to clear these forms of invasive vegetation was a deterrent to further cultivation of the land (see Figure 6.5).

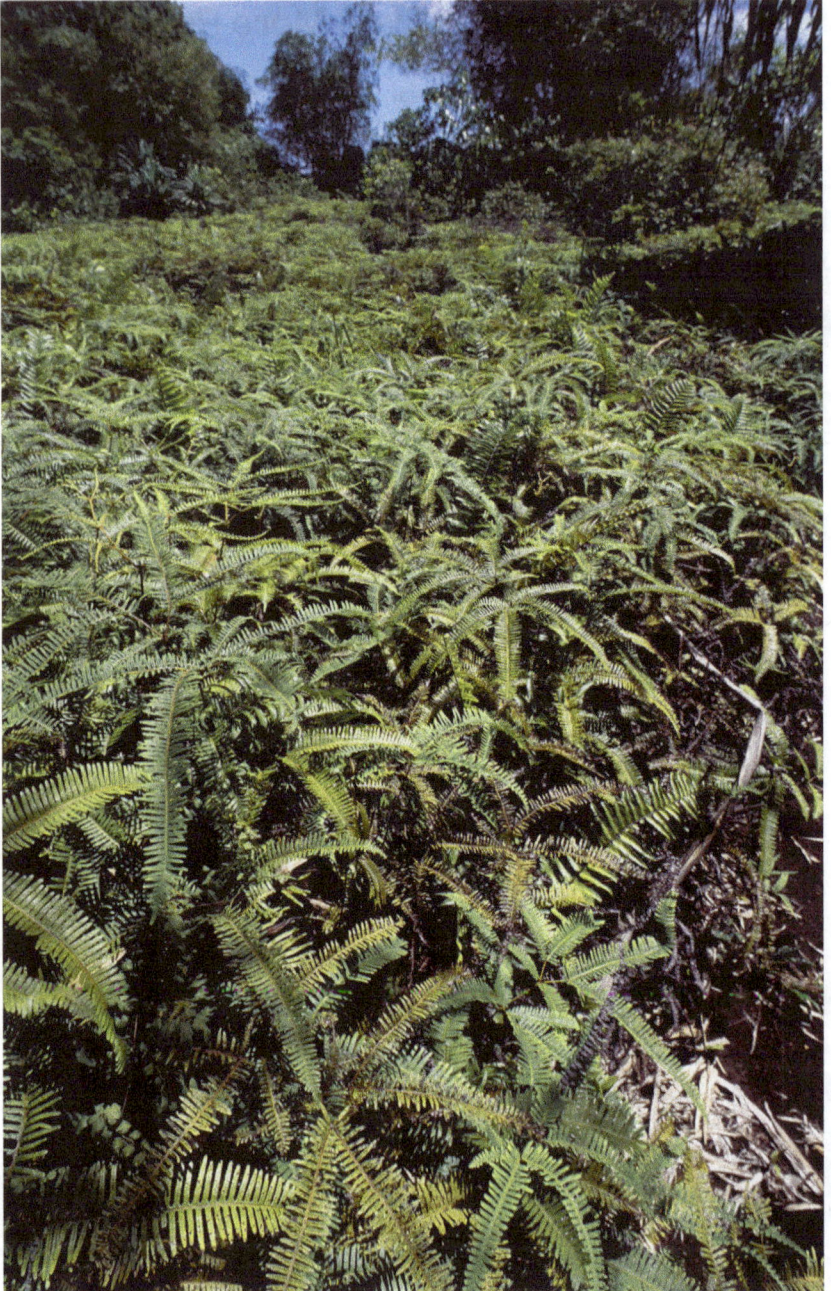

Figure 6.5: Malo ferns on Masahet Island, 2002
Source: Photograph by Simon Foale.

In these early years of the mining operation, people from Malie and Masahet were still trying to deal with the scarcity of cultivable land on their own islands by making gardens in the northeastern part of the main island (Macintyre 2000: 24). By 2003, people from Malie had gardens that covered a single block of land, about 47 hectares in extent, within an undeveloped part of the area leased to the mining company for its town site and migrant workers' accommodation blocks (Macintyre and Foale 2004a: 35). The total amount of land in the mine-affected area that was still available for gardening was evidently shrinking more rapidly than it was on the small islands, even while the resident population in this area was growing at an even faster rate. So, if it were true that 50 per cent of Malie households, and 25 per cent of Masahet households, were now reliant on food obtained from gardens on Aniolam, it was hard to see how this source of supply could be maintained, let alone expanded (ibid.: 21).

In the early years of the mining operation, some of the households on the small islands were still able compensate for the scarcity of gardening land with the wages obtained from their participation in the mining workforce. In 1999, they accounted for 73 (or 31 per cent) of the 234 Lihirians directly employed by the mining company (Macintyre 2000: 59). By 2003, the number had gone up to 91, but this was now only a quarter of the number of Lihirians working directly for the company (Macintyre and Foale 2004a: 72). There would have been other people from the small islands who were employed by the company's local contractors, or who had other jobs in the formal sector workforce, but the authors of the social monitoring reports were unable to discover their number. The 2000 national census counted 472 households in Ihot, so it is reasonable to assume that three-quarters of these households had no direct access to a wage income from employment in the formal sector workforce.

At that juncture, the uneven distribution of cash incomes between the households of Ihot was reflected in the state of their housing. In 1999 there were 88 permanent and 143 semi-permanent houses out of a total of 545 houses on the small islands (Macintyre 2000: 29). 'Semi-permanent' houses were those with corrugated iron roofing and some milled timber woodwork. The remaining 314 houses were made of the 'bush materials', like bamboo and sago, that were partly imported from Aniolam. Some of the housing improvements were ultimately funded by the mining company though an institution known as the Village Development Scheme, but others were based on the incomes of the occupants.[12] The proportion of bush-material houses was notably higher on Mahur than it was on Masahet and Malie, which reflected the lower average cash incomes on the most remote of the three islands. In Putput village, adjacent to the mine site, there were no bush-material houses at all.

Figure 6.6: Ceremonial men's house on Malie Island, 2007
Source: Photograph by Nicholas Bainton.

In 2001, the disparities between the three islands of Ihot were given a further boost when the people of Malie mounted a successful campaign for their island to be included in the 'mine-affected area' because of the damage that it was supposedly suffering from the operation of the mine (Macintyre and Foale 2002: 142–7).[13] This meant that they would be entitled to new forms of compensation, including further improvements to their housing stock, which would not be made available to residents of the other two islands. Within a few years, every house on Malie, including the ceremonial men's houses, was constructed from permanent materials (see Figure 6.6). This has

12   In 1998, the Malie islanders opted to take the cost of building six of these houses as a cash payment so that they could build 12 houses themselves at half the unit cost. The Masahet islanders expressed a similar preference, but were still waiting for the money to arrive (Filer and Mandie-Filer 1998: 15).

13   The Lihir Mining Area Landowners Association, which is the body representing the lease area landowners, already recognised Malie Island as part of the mine-affected area at the start of mining operations. That was not because of the environmental impact of the mining operation on the island itself, but because the islanders traditionally harvested subsistence resources in the area surrounding the new town site (Filer and Mandie-Filer 1998: 3).

increased the demand for residential land, as each new household expects to build a permanent house that requires an area of at least 90 m²—an area roughly equivalent to half a tennis court. As a result, the residential zone on Malie is very tightly packed, and some people have started to build their houses in areas formerly reserved for gardens and tree crops. The incidence of land scarcity across the three islands was now quite clearly correlated with the extent of people's dependence on the continued operation of the mine to provide the incomes or benefits required to compensate for the pressure that rapid population growth was placing on the local agricultural system.

Left to Their Own Devices

The mining company terminated the social monitoring program in 2004. This meant that the collection of evidence about the social and economic changes taking place in different parts of the Lihir island group became the responsibility of company staff who were already fully engaged in the management of 'community affairs'. In any case, the company had not previously shown much inclination to act on the recommendations of its consultants, including those directed towards the peculiar problems of the small island population. The people of Ihot have thus been left to devise their own solutions to these problems.

When Macintyre produced the first graph showing the 'land needs' of the inhabitants of each of the small islands as compared with those of the main island, she observed that

> [t]he problem of overpopulation on Malie and Masahet is rapidly approaching the point where the environmental issues relating to long-term economic sustainability will constitute a crisis. This is not recognised by most Lihirians and those who do acknowledge it do not see reduction in the birthrate as a solution.
>
> (Macintyre 2000: 19)

As predicted in the earlier social impact studies, the local influence of the Catholic Church, combined with the subordinate status of women in Lihirian society, discouraged the implementation of birth control initiatives, and this in turn had contributed to ongoing population growth on the smaller islands. At the same time, community members have not consistently raised such issues with the company or the government because their attention has mostly been absorbed by the political process of negotiating the unequal distribution of economic benefits and environmental impacts associated with the mining operation (Macintyre and Foale 2004b).

One thing that did survive the end of the social monitoring program was a demographic monitoring program, the results of which were stored on a database, known as the Village Population System (VPS), which John Burton had installed in the company's computers before the start of mine construction (Burton 1994, 2007). This was intended to reassure local community leaders that company managers would be able to tell the difference between genuine Lihirians and migrants from other parts of PNG when allocating jobs in the mining workforce (Bainton 2017: 322–3). By the end of 2014, the VPS database showed the extent of the change that had taken place in the Lihirian rural village population on each of the four islands in the Lihir group since 1980 (see Table 6.5).[14] An updated version of the 'land needs' graph would show that the Masahet islanders had gone well past the threshold of sustainability by 2014, and even the Mahur islanders would now be approaching it (Bainton et al. 2022: Figure 2). However, it is evident that the overall rate of population growth on each of the three small islands since 1980 has been rather oddly correlated with the extent to which they are short of cultivable land, as well as being correlated with their proximity (or ease of travel) to the main island. On the other hand, the rate of population growth on the main island has been considerably higher than on any of the three small islands. And since previous calculations suggest that there is more cultivable land per head of population on the main island than on any of the three small islands, there appears to be no simple relationship between population growth rates and the degree of population pressure on local agricultural systems.

Table 6.5: Lihirian rural village populations on the four islands in 1980 and 2014

Island	1980			2014			% change		
	M	F	All	M	F	All	M	F	All
Aniolam	1,915	1,724	3,639	6,690	6,267	12,957	249.3	263.5	256.1
Malie	130	135	265	396	359	755	204.6	165.9	184.9
Masahet	439	402	841	904	850	1,754	105.9	111.4	108.6
Mahur	315	311	626	600	536	1,136	90.5	72.3	81.5
TOTAL	2,799	2,572	5,371	8,590	8,012	16,602	206.9	211.5	209.1

Source: 1980 national census and 2014 Village Population System.

14　These figures do not include the number of Lihirians resident in the three 'rural non-villages' on the main island.

A comparison of the numbers in Tables 6.4 and 6.5 indicates that the excess of males over females, already evident in 1958, was still present in 2014. It also reveals that the population growth rate has been higher on the main island than on any of the small islands throughout this longer period. On the other hand, differences in the rate of growth between the three small islands appear to have changed over the course of this period. While Malie had the lowest growth rate (27.4 per cent) in the period from 1958 to 1980, it had the highest growth rate (184.9 per cent) in the period from 1980 to 2014. The reverse was the case in Mahur: a relatively high growth rate (48.3 per cent) in the earlier period was followed by a relatively lower growth rate (81.5 per cent) in the later period.

These discrepancies cannot simply be due to differences in fertility and mortality rates on different islands. In the earlier period, as we have seen, the lower growth rate in Ihot, as compared to Aniolam, was partly due to a higher rate of emigration from the small islands. The question then is whether migration also helps to explain the discrepancies observed in the more recent period.

As predicted in the original social impact studies, most of the Lihirians exiled in other parts of PNG had returned to the islands by the time that mining operations began in 1997. This was reflected in the 2000 national census, which counted a rural village population of 10,924—more than twice the number counted in the 1980 census. But development of the mining project then began to have an independent effect on the distribution of the local population. This is immediately evident if we compare the rate of population growth in the northeastern part of Aniolam that was designated as the 'mine-affected area' in the original development agreements with the rate of population growth in the rest of the main island (see Table 6.6).

Table 6.6: Lihirian rural village populations in two parts of Aniolam in 1980 and 2014

Area	1980			2014			% change		
	M	F	All	M	F	All	M	F	All
'Affected'	545	466	1,011	2,371	2,179	4,550	335.0	367.6	350.0
'Unaffected'	1,370	1,258	2,628	4,319	4,088	8,407	215.3	225.0	219.9
TOTAL	1,915	1,724	3,639	6,690	6,267	12,957	249.3	263.5	256.1

Source: 1980 national census and 2014 Village Population System.

Here we can see that the population of the 'affected' villages grew by 350 per cent between 1980 and 2014, while the population of the 'unaffected' villages grew by 220 per cent. The growth rate in the unaffected villages was much the same as the growth rate for the whole island group, but was much higher than the growth rate for the small islands of Ihot (see Table 6.5). It therefore seems reasonable to infer that a substantial number of people from Ihot have taken up residence in the affected area on Aniolam since the start of mining operations, while very few people from the unaffected parts of Aniolam have done so. In which case, it would appear that the advent of the mining project has created a new form of symbiosis between the main island and the small islands that reaches beyond the practices already documented in the original social impact studies.

This new form of symbiosis must also be partly shaped by the fact that the population living in this mine-affected area is actually much higher than might be inferred from the figures shown in Table 6.6. That is because a very substantial number of undocumented migrants from other parts of PNG have settled in this area under the sponsorship of those same community leaders and local landowners who insisted on their exclusion from the VPS database (Filer and Mandie-Filer 1998). Given that these people occupy a sort of liminal social space in which their presence is both recognised and denied, it has been very difficult to count them, let alone to determine what they do for a living. However, most of the adult members of this migrant community seem to be employed, more or less formally, by Lihirian patrons who derive most of their own income from their relationship to the mining company—whether as recipients of wages, royalties, business contracts or other benefits distributed under the terms of the development agreements (Bainton and Macintyre 2013; Bainton 2017). If we add the workers formally and directly employed by the mining company, who are either permanent residents of the township or temporary (commuting) residents of the nearby accommodation blocks, there could have been as many as 10,000 people resident in this mine-affected area in 2014.

If people still living on Malie and Masahet islands are still attempting to cultivate food crops in what is now the mine-affected area on Aniolam, which is the only part of the main island where they used to make extra gardens in the past, their task is now a great deal more formidable. That is because there is a lot more competition for access to a rapidly diminishing area of cultivable land in this part of the main island, just as there is on the two small islands, and some of that competition would be coming from relatives who have recently relocated themselves and their families from the small islands to the main one. However, the development of the mining economy has also reduced the relative importance of food and cash crop cultivation as a livelihood strategy for a substantial proportion of the Lihirian population, especially those living in the mine-affected area, which now includes Malie Island. It is therefore necessary to consider some of the other ways in which this new mining economy has influenced the pattern of inter-island relationships.

Time–Space Distanciation and the Intensification of Custom

Since Macintyre and Foale last reported on the number of mine workers who come from the small islands, the absolute number of employees from Ihot has continued to grow, even though the proportion of Ihot residents in the total Lihirian mining workforce has diminished. This generally has more do with the company's commitment to 'local Lihirian' employment as a condition of the benefit-sharing agreement than any explicit strategy to offset the pressures facing small island communities. For example, in 2011 the mining company directly employed some 2,212 people. Lihirians made up 33 per cent of the workforce. About 18 per cent of these Lihirians came from Ihot, which represented a significant drop from 1999 when Ihot residents made up 31 per cent of the workforce. In 2011 Ihot workers received approximately 4.3 million kina in salaries and wages for that year alone (Bainton 2013). Given the increase in contracting opportunities for local landowner businesses (Bainton and Jackson 2020), it is likely that a comparable number of Ihot residents have been employed by these local companies. Although a much smaller number of people from Ihot have established businesses that can contract with the mining operation, these businesses still pull a considerable flow of income to the small islands.

Overall, salaries and wages represent the most significant source of income and are spread more evenly across the four islands compared with other mine-derived income streams. Mineral royalties constitute a discernibly smaller portion of these flows.[15] The distribution of royalty payments across Lihirian society reflects the ownership patterns of lease land areas in terms of geographical

---

15   The royalty rate is set at 2 per cent of gross revenue from gold sales. The royalties are split as follows: New Ireland Province (50 per cent); Nimamar (Lihir) Local Level Government (30 per cent); Special Mining Lease Landowners (20 per cent). A lack of public reporting has made it difficult to discern how the Local Level Government and the Provincial Government have allocated and spent their portions of the royalty payments.

proximity and inheritance within matrilineal descent groups. The villages that are closest to the mine contain more lease area landowners, and these villages have consequently received the largest share of royalty payments. In 2011 the company paid out 46.2 million kina in royalties. Although 8 million kina was paid directly to lease area landowners, only 72,000 kina was received on the smaller islands, which reflects the smaller number of people on Ihot who can claim rights over parcels of lease land on Aniolam.

The flow of mine-derived income from Aniolam to Ihot points to numerous spatio-temporal changes occurring in Lihir. For the residents of Ihot, time and space have simultaneously shrunk and stretched. Anthony Giddens refers to this as 'time–space distanciation', by which he means the processes whereby societies are stretched over longer or shorter spans of time and space (1995: 90). On one hand we find economic and technological changes have created a 'compression' effect, dramatically reducing the time needed to access the islands or to communicate with people living on Aniolam and beyond. On the other hand, these changes have 'stretched' social relations and accelerated time as mine-derived incomes are ploughed into customary activities in ways that expand the network of people embedded in these relations of reciprocity and increase the frequency and intensity of these events. These twin processes shape the pressures facing the residents of Ihot and how they have responded to them. Let us explain.

Following the earlier recommendation for a ferry service to link the smaller islands to the main island, the mining company has operated a fleet of fibreglass banana boats to ferry Ihot workers to the mine site (see Figure 6.7). The majority of Ihot workers make the long (and wet) commute daily, while a smaller fraction resides in the mining camp if they have a 'special case' for accommodation; others make arrangements with relatives in affected-area villages, which is one of the ways that a growing number of Ihot people have taken up residence in the affected area on Aniolam. Access to more cash has also resulted in a greater number of privately owned banana boats across all four islands. This has increased social connectivity between the residents of Aniolam and Ihot. It has also made it possible for Ihot residents to access services, and for some residents to regularly make subsistence gardens, in northeastern Aniolam. Access to mobile telecommunications has also dramatically improved the situation on the smaller islands and provided enhanced connection to friends and relatives on Aniolam and in the outside world, as it has elsewhere throughout the Pacific Islands (Foster and Horst 2018). Mobile technology has been especially important for enhanced safety at sea and helping to locate travellers that find themselves adrift.

Figure 6.7: Fibreglass banana boats at Mahur, 2016
Source: Photograph by Nicholas Bainton.

Despite the improvements to life on the smaller islands, these changes have not entirely offset pressures on land and natural resources. In fact, the expansion of the mining economy has paradoxically contributed to other factors (besides population growth) that have intensified these pressures. As more Lihirians are drawn into the cash economy and substitute store goods for garden goods, this has resulted in the gradual loss of gardening skills and knowledge. This shift has most likely decreased horticultural efficiency and increased the amount of land required per head of population. Very few Lihirians have adopted agricultural intensification techniques, on Aniolam or Ihot, and efforts by the mining company, the local-level government and the landowner association to support agricultural extension services throughout Lihir have been inconsistent at best.

A significant portion of the wealth derived from the mining economy has been absorbed into the local ceremonial economy, which revolves around an extended cycle of mortuary feasts (Bainton and Macintyre 2016). As customary activities have flourished—as a kind of cultural response to the effects of resource extraction—this has simultaneously increased the pressure on local resources and increased the number of disputes concerning the inheritance of land on the smaller islands and Aniolam. The demand for extra pigs and yams as key items of exchange within that ceremonial economy can be considered as a key driver of land pressure (and ecosystem change) in its own right, because pigs and yams both make their own demands on the local food-cropping systems. Ceremonial yam gardens in particular are likely to have a significant impact, because there is much prestige associated with the size of yams presented at feasts, and the most effective means of achieving large yam sizes is to clear new areas of primary forest, since soil fertility tends to be higher here than on land that has been under fallow. Since this agricultural strategy is no longer possible on Malie and Masahet, people on these islands are further reliant upon their relatives on Aniolam for access to large yams. Ceremonial yam gardens tend to be much larger in area than gardens established for domestic subsistence. To meet the expanding needs of this ceremonial cycle, Lihirians have also made a habit of purchasing pigs and garden produce from fellow Lihirians and from other island groups, which has become a mechanism by which Lihirians redistribute a portion of the surplus income obtained from the mining economy. This practice has less to do with conserving scarce resources and is more related to the speeding up (or the compression) of the local feasting cycle, which has generated a requirement to mobilise larger feasts at shorter notice. Ready access to sea transport makes it easier for large groups of people to travel to the smaller islands for customary feasts for shorter periods of time, and for hosts and donors to roam around the local region in search of pigs that can be bought with a combination of cash and shell money for exchange at these events.

Figure 6.8: Customary feast on Mahur Island, unveiling of *mormor* (memorial depicting deceased clan members), 2015
Source: Photograph by Nicholas Bainton.

When rich landowners (some of whom trace their descent to the outer islands) co-host or contribute to extravagant mortuary feasts on Ihot, they direct the flow of mining wealth from Aniolam to the smaller islands. As an indirect effect of mining, this flow of wealth provides collective prestige for communities and clans while simultaneously squeezing individual households: pressures are transmitted across the seas, from one island to the next, through kinship lines as clan members exhaust their resources in order to participate in these moments of ritual excess (see Figure 6.8). Just as it was once recognised that Ihot communities had constructed the most elaborate church buildings, Ihot residents are now recognised for building the most elaborate ceremonial men's houses and hosting the most elaborate feasts. Many Lihirians regard the smaller islands, particularly Masahet and Mahur, as the seat of customary practices and knowledge because they are less proximate to the mine and its impacts and because there has been a stronger commitment to 'strengthen custom' within these communities (Bainton et al. 2011). As a result, local big men on Ihot are seen as the guardians of Lihirian custom, even if their customary feasts and exchanges have become increasingly competitive and reliant upon the mining economy and threaten to spiral out of control. In this respect, there is also something of a paradox in the cash-fuelled expansion of the ceremonial economy and the simultaneous erosion of traditional technical and environmental knowledge.

Conclusion: Crisis, Class and Constraint

In a report written for the PNG Department of Environment and Conservation in 1992, Filer suggested that the negative impacts of the future mining operation could be grouped under two main headings, which he called the 'stratification effect' and the 'demoralisation effect'. The first of these was said to follow from 'the probability that different members of the local community will experience the different aspects of the development process in different forms and degrees, and the process as a whole will therefore give rise to new forms of inequality, division and conflict within the community' (Filer 1992b: 5). The second was said to follow from:

> the probability that the community as a whole will be 'overpowered' by the presence of the project, existing mechanisms of social control will be disrupted and devalued, and local people's respect for 'custom' will progressively be transformed into a frustrating sense of dependency on the project as the source of all their problems and the only source of their solution.

> (ibid.: 5)

In the same report, he suggested that 'custom' was 'the largest single umbrella beneath which the people of Lihir still choose to take shelter in a fairly wide range of circumstances', even if it might prove to be 'an inadequate defence against the stratification effect, and may be further weakened as a consequence' (ibid.: 6).

In a subsequent report to local community leaders in 1994 (which was mostly written in Tok Pisin), Filer listed seven 'social problems' that were likely to arise as a result of the mining operation. These were: (1) 'resource scarcity or over-population'; (2) the 'influx of outsiders'; (3) the 'breakdown of law and order'; (4) the 'unequal distribution of income'; (5) the 'subordination of women'; (6) 'power struggles within the community'; and (7) the 'project dependency syndrome' (Filer 1994: 2). The first six of these problems could be understood as manifestations of the stratification effect, while the last made reference to the demoralisation effect. These predictions have turned out to be fairly accurate, and it is hardly surprising that all these problems have been most acute in what came to be designated as the mine-affected area. But how have they manifested themselves in the relationship between Ihot and Aniolam?

The crisis confronting the outer islands can be understood as a type of cumulative impact, where pressures that were evident before mining started have intensified and transformed over the duration of the mining operation. Although mining wealth has evidently flowed to the residents of Ihot, and helped to improve material living standards, these 'benefits' of extraction are also a source of inequality that amplify social tensions throughout the islands. Class differences have now more or less settled into place. The majority of Lihirians who command control over mine-derived forms of wealth live in the northeastern corner of Aniolam, the mine-affected area. Customary obligations may cut across these differences and provide conduits for the circulation of wealth, but these distributive moments and movements cannot unsettle the structural differences between those people who enjoy privileged and continuing access to mining capital and services, and those people who must content themselves with lesser prospects. For the population on the smaller islands, these differences appear at night as distant shimmering lights of 'development' that are easily seen when they look across the seas to Aniolam. In this regard, both the stratification effect and the demoralisation effect are manifest in Ihot through the feeling of resentment and relative deprivation.

In the long term it remains to be seen whether the wealth produced by the mine will be converted into a form of sustainable development that can support the small island population into the post-mining future. While the process of mine closure will play out over an extended period and may present new opportunities for future planning, current prospects do not look promising. This is despite great efforts by the Lihirian political elite to devise and implement their 'Lihir Sustainable Development Plan' (Bainton 2010; Bainton and Macintyre 2021). In the meantime, a growing number of households will have to contend with the effects of a 'simple reproduction squeeze' (Bernstein 1979) that has gripped the smaller islands. As we have demonstrated, population growth combined with other novel factors makes it harder for some households to access land and resources for subsistence and increases their dependency upon the (finite) mining economy. This squeezing effect takes a variety of historically specific forms, most notably through the proliferation of customary activities and obligations. These ceremonial commitments also act as another anchor that prevents some Ihot households from moving elsewhere to escape the clutches of this reproduction squeeze.

While some Ihot residents have managed to move to Aniolam, the real puzzle is how many of them are doing what to make a living there. The company's monitoring programs have not tracked this trend, and in the absence of such data it is difficult to predict the long-term prospects for these people. On the other hand, very few residents from Ihot or Aniolam have left the islands altogether. For most small island communities, out-migration is the only viable coping strategy. In Torres Strait, for example, patterns of circular migration have emerged, whereby younger residents migrate down to mainland Australia in search of jobs and services that are absent in the islands. Many of these 'emigrants' later return to their island homes to retire. Similar patterns have long been evident throughout PNG. These circular rural–urban–rural movements are shaped by economic drivers and strong attachments to 'place' (McGavin 2016). But as previously observed, mining has given rise to new patterns of rural–rural, or rural–resource migration (Bainton and Banks 2018). Local residents, who may be displaced by the mine or crowded out by in-migrants, rarely want to move to other places out of fear that they will lose their just entitlement to mining benefits. And in the case of Lihir, the Ihot population seems to be fixed to an even greater extent as a result of an extractive–ceremonial complex. In other cases and places, there are not the same attractors.

We envision several possible future scenarios for the communities of Ihot. The long-term post-mining outlook may revolve around increasing pressure on customary arrangements to access land on Aniolam. It is unlikely that many people will purchase land on mainland New Ireland or elsewhere in PNG. While some Lihirians have already purchased properties abroad, in most cases these are 'investments' that generate an income that allows them to remain in Lihir. This trend may increase among wealthier Lihirians. As income-generating opportunities decline in the post-mining era, becoming landlords in other places may be one of the most viable economic opportunities available to Lihirians. The option to invest in other places is probably more realistic for lease area landowners with access to money, and not likely to be an option that many people from Ihot can afford. At present, very few Lihirians permanently reside away from the islands, and it is difficult to predict whether this will change in the future. For most Lihirians, the balance of attractions is presently greater than the balance of impacts—or there are other structural or cultural reasons that prevent them from moving elsewhere. This balance may shift with mine closure, and we may witness a more mobile Lihirian population. This attraction, and the balance itself, is a major driver of pressures. Despite the emergent challenges of mine-induced land scarcity, the balance of attractors means that the vast majority of Lihirians have elected to remain in their group of islands. This in turn highlights the land and services paradox: as services become increasingly viable as the population grows (as the cost per capita decreases), land—itself a core provider of services—has a finite capacity.

Although the pressures that are currently faced by the Ihot population are not climate-induced, sea level rise and coastal inundation and erosion are being experienced in some locations, especially on Malie. However, future climate changes will exacerbate other pressures and existing vulnerability. If there are longer and more frequent droughts this will make life more intolerable on Ihot and increase dependency on the mining economy for subsistence. Climate effects are likely to be uneven, exposing the inequalities between the people living on the main island with more resources at hand, and those living on the smaller islands with more limited options to choose from.

In the end, the solutions to land pressure on small islands are limited, and the income streams that supplement subsistence-based livelihoods are contingent upon the operation of the mine. For this reason, it is worth recalling the conclusions reached by the authors of the original social impact assessments for the mine: if no steps are taken to address the crisis confronting the outer islands, then the Ihot population will have to live with the fact that there is no way of supporting their projected population with the present system of subsistence. The population of the outer islands will not simply be unsupportable in the manner to which mining will have accustomed them but will be unsupportable in *any* reasonable manner. A steadily growing population of Ihot may have no option but to migrate in search of an alternative livelihood. The question is, where will they go?

References

Anon, 2019. 'Lihir's $719m in Royalties.' PNGReport blog post, 10 January 2019. www.pngreport.com/mining/news/1353980/lihirs-usd719m-in-royalties (site discontinued).

Bainton, N., 2010. *The Lihir Destiny: Cultural Responses to Mining in Melanesia.* Canberra: ANU Press (Asia-Pacific Environment Monograph 5). doi.org/10.22459/LD.10.2010

——, 2013. 'Lihir Social and Economic Impact Assessment.' Unpublished report to Newcrest Mining Ltd.

——, 2017. 'Migrants, Labourers and Landowners at the Lihir Gold Mine, Papua New Guinea.' In C. Filer and P.-Y. Le Meur (eds), *Large-Scale Mines and Local-Level Politics: Between New Caledonia and Papua New Guinea.* Canberra: ANU Press (Asia-Pacific Environment Monograph 12). doi.org/10.22459/LMLP.10.2017.11

Bainton, N., C. Ballard, K. Gillespie and N. Hall, 2011. 'Stepping Stones across the Lihir Islands: Developing Cultural Heritage Management in the Context of a Gold-Mining Operation.' *International Journal of Cultural Property* 18: 81–110. doi.org/10.1017/S0940739111000087

Bainton, N. and G. Banks, 2018. 'Land and Access: A Framework for Analysing Mining, Migration and Development in Melanesia.' *Sustainable Development* 26: 450–460. doi.org/10.1002/sd.1890

Bainton, N., J. Burton and J.R. Owen, 2022. 'Land Relations, Resource Extraction and Displacement Effects in Island Papua New Guinea.' *Journal of Peasant Studies* 49: 1295–1315. doi.org/10.1080/03066150.2021.1928086

Bainton, N. and R.T. Jackson, 2020. 'Adding and Sustaining Benefits: Large-Scale Mining and Landowner Business Development in Papua New Guinea.' *Extractive Industries and Society* 7: 366–375. doi.org/10.1016/j.exis.2019.10.005

Bainton, N. and M. Macintyre, 2013. '"My Land, My Work": Business Development and Large-Scale Mining in Papua New Guinea.' *Research in Economic Anthropology* 33: 137–163. doi.org/10.1108/S0190-1281(2013)0000033008

——, 2016. 'Mining Riches and Mortuary Ritual in Island Melanesia.' In D. Lipset and E. Silverman (eds), *Mortuary Dialogues: Death Rites and the Reproduction of Moral Community in Pacific Modernities.* Oxford: Berghahn Books.

——, 2021. 'Being Like a State: How Large-Scale Mining Companies Assume Government Roles in Papua New Guinea.' In N.A. Bainton and E.E. Skrzypek (eds), *The Absent Presence of the State in Large-Scale Resource Extraction Projects.* Canberra: ANU Press (Asia-Pacific Environment Monograph 15). doi.org/10.22459/AP.2021.04

Barnett, J. and J. Campbell, 2010. *Climate Change and Small Island States: Power, Knowledge and the South Pacific.* London: Routledge. doi.org/10.4324/97818 49774895

Bernstein, H., 1979. 'African Peasantries: A Theoretical Framework.' *Journal of Peasant Studies* 6: 421–443. doi.org/10.1080/03066157908438084

Bonnell, S., 1992. 'Lihir Food Garden Survey, 28 September to 28 October 1992.' Unpublished report to Lihir Management Company.

Burton, J.E., 1994. 'The Lihir VPS Database: A Tool for Human Resources Planning, Social Development Planning and Social Monitoring.' Port Moresby and Canberra: Unisearch PNG and Pacific Social Mapping for Lihir Management Company and PNG Department of Mining and Petroleum.

——, 2007. 'The Anthropology of Personal Identity: Intellectual Property Rights Issues in Papua New Guinea, West Papua and Australia.' *Australian Journal of Anthropology* 18: 40–55. doi.org/10.1111/j.1835-9310.2007.tb00076.x

Clay, B.J., 1986. *Mandak Realities: Person and Power in Central New Ireland.* New Brunswick: Rutgers University Press.

Connell, J., 2015. 'Vulnerable Islands: Climate Change, Tectonic Change, and Changing Livelihoods in the Western Pacific.' *The Contemporary Pacific* 27: 1–36. doi.org/10.1353/cp.2015.0014

Filer, C., 1992a. 'The Lihir Hamlet/Hausboi Survey: Interim Report.' Waigani: Unisearch PNG Pty Ltd for Kennecott Explorations (Australia) and PNG Department of Mining and Petroleum.

——, 1992b. 'Lihir Project Social Impact Mitigation: Issues and Approaches.' Waigani: Unisearch PNG Pty Ltd for PNG Department of Environment and Conservation.

——, 1994. 'Sosel Impakt Bilong Lihir Gol Main: Ripot i Go Long Pipal Bilong Lihir.' Unpublished report to Lihir Mining Area Landowners Association.

——, 1995. 'Participation, Governance and Social Impact: The Planning of the Lihir Gold Mine.' In D. Denoon et al. (eds), *Mining and Mineral Resource Policy Issues in Asia-Pacific: Prospects for the 21st Century*. Canberra: The Australian National University, Research School of Pacific and Asian Studies, Division of Pacific and Asian History.

Filer, C. and R. Jackson, 1986. *The Social and Economic Impact of a Gold Mine on Lihir*. Port Moresby: Department of Minerals and Energy, Lihir Liaison Committee.

Filer, C.S. and R.T. Jackson, 1989. *The Social and Economic Impact of a Gold Mine on Lihir: Revised and Expanded*. Port Moresby: Department of Minerals and Energy, Lihir Liaison Committee (2 volumes).

Filer, C. and A. Mandie-Filer, 1998. 'Lihirian Perspectives on the Social and Environmental Aspects of the Lihir Gold Mine.' Unpublished report to the World Bank.

Foster, R.J. and H.A. Horst, 2018. *The Moral Economy of Mobile Phones: Pacific Islands Perspectives*. Canberra: ANU Press. doi.org/10.22459/MEMP.05.2018

George, M. and D. Lewis, 1985. 'Maritime Trade and Traditional Exchange in the Bismarck Archipelago.' In J. Allen (ed.), *Lapita Homeland Project: Report of the 1985 Field Season*. Unpublished report.

Giddens, A., 1995. *A Contemporary Critique of Historical Materialism*. London: Macmillan Press.

Gormley, P.L., 1968. 'Lihir Island: Report of Survey of Present Arrangements for Marketing Native Produced Copra.' Namatanai: Territory of Papua New Guinea, Department of Trade and Industry.

Groves, W.C., 1935. 'Tabar To-Day: A Study of a Melanesian Community in Contact with Alien Non-Primitive Cultural Forces.' *Oceania* 5: 346–360. doi.org/10.1002/j.1834-4461.1935.tb00156.x

Hemer, S.R., 2011. 'Local, Regional and Worldly Interconnections: The Catholic and United Churches in Lihir, Papua New Guinea.' *Asia Pacific Journal of Anthropology* 12: 60–73. doi.org/10.1080/14442213.2010.535844

——, 2016. 'Emplacement and Resistance: Social and Political Complexities in Development-Induced Displacement in Papua New Guinea.' *The Australian Journal of Anthropology* 27: 279–297. doi.org/10.1111/taja.12142

Hide, R.L., R.M. Bourke, B.J. Allen et al., 1996. 'New Ireland Province: Text Summaries, Maps, Code Lists and Village Identification.' Canberra: The Australian National University, Research School of Pacific and Asian Studies, Department of Human Geography (Agricultural Systems of Papua New Guinea Working Paper 17).

Macintyre, M., 2000. 'Social and Economic Impact Study: Lihir 1999.' Unpublished report to Lihir Management Company.

Macintyre, M. and S. Foale, 2001. 'Social and Economic Impact Study: Lihir 2000.' Unpublished report to Lihir Management Company.

——, 2002. 'Social and Economic Impact Study: Lihir 2001.' Unpublished report to Lihir Management Company.

——, 2004a. 'Social and Economic Impact Study: Lihir 2003.' Unpublished report to Lihir Management Company.

——, 2004b. 'Global Imperatives and Local Desires: Competing Economic and Environmental Interests in Melanesian Communities.' In V.S. Lockwood (ed.), *Globalization and Culture Change in the Pacific Islands*. Upper Saddle River: Pearson Prentice Hall.

May, R.J. (ed.), 1982. *Micronationalist Movements in Melanesia*. Canberra: Research School of Pacific Studies, Department of Political and Social Change (Monograph 1).

McGavin, K., 2016. 'Where Do You Belong? Identity, New Guinea Islanders, and the Power of *Peles*.' *Oceania* 86: 57–74. doi.org/10.1002/ocea.5112

Neuhaus, K., 1932. 'Ich Habe Mich Verändert: Plauderei des Missionars von Lihir, P.K. Neuhaus, M.S.C.' *Hiltruper Monatshefte* 49: 336–342.

——, 2015 [1954]. *Grammar of the Lihir Language of New Ireland, Papua New Guinea* (transl. S. Ziegler). Boroko: Institute of Papua New Guinea Studies and Lihir Cultural Heritage Association.

Taufa, T., G. Day and V. Mea, 1991. 'Base-Line Health Survey of Lihir Islanders, April 1991.' Hawthorn: NSR Environmental Consultants Pty Ltd.

7

Conclusion

Colin Filer and Simon Foale

When we embarked on our adventure with the Sub-Global Working Group of the Millennium Ecosystem Assessment in 2002, we already had a mandate to think about small islands in peril or under pressure. However, it needs to be borne in mind that our mandate was generated by the proximity of a specific group of small island communities to the location of marine biodiversity values that were the principal focus of concern to the other parties engaged in a project funded by the Global Environment Facility. So the islanders had to be encountered as a group of actors whose existing livelihood strategies were more or less of a threat to these biodiversity values, and who might or might not be persuaded to adjust these strategies in order to reduce the threat that they posed.

As members of the Sub-Global Working Group, we were encouraged to think of these island communities and their marine environments as 'social–ecological systems' of a certain type. However, when we were obliged to broaden the scope of our vision to include a much larger number of communities, we soon began to doubt whether there was any way to assign each community to a certain type of system. The final report of the Sub-Global Working Group included an assertion that local communities are located at the bottom of a hierarchy of scales at which an ecosystem assessment could be undertaken, and that members of these communities are more or less able to deal with the problem of environmental degradation through the application of their own traditional knowledge and resource management practices, depending on the scale at which the problem is being created (Folke et al. 2005). This way of dealing with the question of scale and the interaction of different 'knowledge systems' is a common feature of the academic work produced by members of the Resilience Alliance (www. resalliance.org), and featured in a separate contribution to the Millennium Ecosystem Assessment (Reid et al. 2006). Despite the widespread adoption of this conceptual framework, we were still troubled by the assumption that social institutions and knowledge systems could be attached to ecosystems distinguished as spatial polygons at a particular geographical scale and then construed as instruments for the management—or mismanagement—of the services provided by those ecosystems to human consumers. Our own attempt to think this way, as illustrated in Figure 3.1, made us feel as if we were revisiting the form of cultural ecology propounded by the American anthropologist Julian Steward (1955), in which local 'cultures' are treated as 'adaptations' to local environments.

With that point in mind, the answer to the question implied in the title of this volume should now be obvious. If we think of island ecosystems as ecosystems of a certain type defined at a particular scale, then the size of an island has very little relationship to the lives or livelihoods of the people who live on it or the community of which it is a part. The size of an island is simply something that is easy to measure. The act of classification that was prompted by the idea that some small island communities are 'in peril' because their islands are so small is an act that naturally leads to a recognition that size in itself is not a very important element in their material conditions of existence or the state of the terrestrial and marine ecosystems from which they obtain the 'services' that partially sustain them. Island size is just a vantage point from which to examine the way that people deal with the different 'pressures' that shape their livelihoods, including those that entail some form of environmental degradation. In this respect, there would be no point in attempting to construct a representative sample of small island communities or ecosystems in a specific region because the sampling frame would contain too many variables. As anthropologists, we can only offer a partial vision of the range of variation, recognising that much of that variation consists of things that cannot be measured at all.

Among the things that can be measured, it would seem that remoteness, altitude and population density matter more than island size as material conditions of existence, or as what are called 'states' in the pressure–state– response model of social and environmental change. But the way in which they matter varies with the nature of the pressures and the responses. Population density matters when there are high rates of population growth; altitude matters when sea levels are rising; and remoteness matters when islanders are dependent on the consumption of things that have to be imported from somewhere else. However, we are not convinced that the relationship between the three components of this model can be grasped on the assumption that islanders are responsible for the responses that they make to the pressures that they experience. In other words, we are sceptical of the assumption that a specific type of social–ecological system possesses more or less of the qualities known as 'resilience' or 'sustainability' because of the choices made by the people who belong to that system as opposed to those who act on it from the 'outside' (Nadasy 2007; Hornborg 2009). The fact that we have chosen to focus our attention on what goes on in a specific type of place should not prevent us from understanding that what goes on in each place is an effect of the unequal distribution of power between the people associated with different places in different ways.

Of course, anthropologists can also be implicated in this unequal distribution of power. Some of our colleagues, inspired by the work of Epeli Hau'ofa (1994), might argue that we are guilty of 'belittling' the capacity of islanders to improve their own livelihoods, simply by choosing to focus our attention on the small size of the islands on which they live or with which they identify themselves. Some might say that we should either have celebrated the capacity of islanders to be empowered by the revival of traditional forms of resource management (D'Arcy and Kuan 2023), or else paid more attention to the way that other outsiders (or 'outlanders') have misrepresented small islands in imaginative acts of neo-colonial dispossession (Jolly 2007; Alexeyeff and McDonnell 2018), or else allowed the subjects of our own inquiry to speak for themselves instead of casting ethnographic judgment on their lives (Wesley-Smith 2016; Fair 2020). Our defence would be that this is not a book about the contest between alternative ontologies, epistemologies or ideologies. While we acknowledge the shortcomings of the cartographic lens through which islands make their appearance in official statistics or the imaginations of foreign observers, the contributions to this volume are not intended to assess the current state of 'traditional environmental knowledge' in small island communities. For better or worse, most of the islanders whose voices can be heard in these portraits still sound like they are calling out for some form of 'development', or an improvement in their material conditions of existence, not wishing that government officials, foreign investors or boatloads of tourists would simply go away and leave them to their own devices. But the outlandish presence or absence clearly varies a great deal, both in form and intensity, so it is hard to conclude that all these voices are singing the same song.

At one level or scale, which is a regional scale, there is a sense in which all Pacific Islanders *are* now singing the same song, which is a song about climate change. In the 'small island states' of the Pacific Island region, excluding Papua New Guinea (PNG), nearly all communities are coastal communities, and any difference in the size of the islands that they inhabit can reasonably be represented as a divisive distraction from the political imperative of challenging a form of 'development' that will sooner or later outweigh all the other pressures to which these states and communities must find a response. In these circumstances, Hauʻofaʼs intellectual legacy is represented by groups of activists like the Pacific Climate Warriors, whose actions 'lay the foundation for a Pacific-based counter-discourse that challenges disempowering discourses of drowning islands and helpless Islanders' (Fair 2020: 347). But this kind of 'counter-discourse' is almost entirely absent from the narratives contained in the present volume, simply because we have chosen to focus our attention on communities and ecosystems distinguished at a sub-national—and even microcosmic—scale. If one of these entities had been an atoll formation, like the Kilinailau community identified with the Carteret Islands, then the spectre of climate change would have loomed a lot larger. But atoll communities represent a very small fraction of PNG's small island communities, just as small island communities represent a very small fraction of the country's total population, and even a small fraction of its costal population. That is why the voice of the Carteret Islanders or their representatives makes a louder noise in the sphere of international relations than it does at the sub-national scale where we have situated our analysis. At this finer scale, we should not expect the members or representatives of small island communities to have a distinctive political voice for the simple reason that their material conditions of existence and the factors that influence these conditions are themselves so highly variable. And that is not just the case in PNG, which can hardly be described as a 'small island state', but also in the other states that belong to the Melanesian Spearhead Group.

The case studies presented in this volume might be compared to those presented in a recent World Bank study of the relationship between 'main islands' and 'outer islands' in the small island states of the Pacific Island region (Utz 2021). The authors of this study acknowledge that outer islands exhibit a huge range of variation in size, altitude, population density and relative isolation, and also recognise that this range of variation is hard to measure with national census data. However, they do their best to come up with a statistical assessment of the relationship between rates of out-migration and what they call an index of 'remoteness'. They are not surprised to find that 'migrants move away from more remote localities toward less remote localities' (Utz 2021: 39), and that 'migration helps keep populations on outer islands stable and mitigates pressures on fragile ecosystems that could arise from expanding outer island populations' (ibid.: 53). To lay the ghost of Epeli Hauʻofa, they go on to declare that:

> The 'Sea of Islands' perspective may contain some hints about ways forward. This implies a system that balances traditional norms and strong relationships that should be maintained in outer islands with the importance of boosting connectivity for migration and agglomeration benefits on main islands.

(ibid.: 54)

Regardless of the jargon, the key recommendation of this study is that Pacific Island governments should avoid the provision of subsidies for unprofitable economic activities on outer islands and do what they can to facilitate various forms of migration.

Unlike the authors of the World Bank study, the contributors to this volume would not expect their observations to make any difference to the practices of government agencies. Indeed, as anthropologists, we are inclined to doubt whether government agencies are willing or able to follow the advice provided by the aid industry. Insofar as the state makes an appearance in this volume, it does so in ways that have nothing to do with the World Bank's observations about connectivity or migration, and in ways that do not seem to be accepted or appreciated by the members of small island communities. At the same time, we find it rather curious that the case studies of outer island livelihoods presented in the World Bank study are not based on the writings of anthropologists or other social scientists who have taken the trouble to conduct fieldwork in island communities, but are instead drawn from the pages of Wikipedia. This is most likely due to the fact that anthropologists were not directly engaged in the study, and the authors did not have time to sift through a pile of detailed ethnographic accounts in a search for information that would have some direct bearing on their portrait of contemporary island livelihoods. But it could also reflect the fact that many anthropologists have shifted their own attention from the production of such accounts to the critical interrogation of the powers exercised or abused by national governments, foreign investors or members of the 'donor community', including the World Bank (Jolly 2007).

Many years ago, the geographer Murray Chapman pointed to a different source of weakness in the kind of analysis presented in the World Bank study, which can also be read as a source of weakness in the conceptual framework of the Millennium Ecosystem Assessment and the Resilience Alliance. This is the idea that a community's 'responses' to a change in its material conditions of existence, or the supply of ecosystem services from its immediate neighbourhood, can be characterised by some general statement about patterns of migration or circulation.

> Metaphors such as 'rural–urban drift' and 'circulation,' or technical terms like 'emigration' and 'depopulation' that evoke powerful images, do not ipso facto convey the contemporary ebb and flow of Pacific Island movement, nor its inherently volatile and ambiguous character.

(Chapman 1991: 265)

Chapman was attempting to deconstruct the dualistic metaphors whereby Western scholars, including anthropologists, had sought to understand the movement of Pacific Islanders between different kinds of places—or, if you like, between different types of social–ecological systems distinguished at a certain scale. Nowadays, some anthropologists and geographers might say that these unfounded dualistic metaphors include the contrast between land and water, nature and culture, or society and environment. The contributions to this volume do not go quite so far. Instead, they allow for the existence of small island communities and ecosystems whose local members or managers can make their own distinctions between such things, but without making the assumption that there is a single model or conceptual framework that can make sense of their behaviour, let alone comprehend their interaction with all the other actors who exert some influence over their lives.

References

Alexeyeff, K. and S. McDonnell, 2018. 'Whose Paradise? Encounter, Exchange, and Exploitation.' *Contemporary Pacific* 30: 269–295. doi.org/10.1353/cp.2018.0028

Chapman, M., 1991. 'Pacific Island Movement and Socioeconomic Change: Metaphors of Misunderstanding.' *Population and Development Review* 17: 263–292. doi.org/10.2307/1973731

D'Arcy, P. and D.D.D.-W. Kuan (eds), 2023. *Islands of Hope: Indigenous Resource Management in a Changing Pacific*. Canberra: ANU Press. doi.org/10.22459/IH.2023

Fair, H., 2020. 'Their Sea of Islands? Pacific Climate Warriors, Oceanic Identities, and World Enlargement.' *Contemporary Pacific* 32: 341–369. doi.org/10.1353/cp.2020.0033

Folke, C., C. Fabricius and others, 2005. 'Communities, Ecosystems, and Livelihoods.' In D. Capistrano, C. Samper, M.J. Lee and C. Raudsepp-Hearne (eds), *Ecosystems and Human Well-Being: Multiscale Assessments, Volume 4.* Washington (DC): Island Press.

Hauʻofa, E., 1994. 'Our Sea of Islands.' *Contemporary Pacific* 6: 148–161.

Hornborg, A., 2009. 'Zero-Sum World: Challenges in Conceptualizing Environmental Load Displacement and Ecologically Unequal Exchange in the World-System.' *International Journal of Comparative Sociology* 50: 237–262. doi.org/10.1177/0020715209105141

Jolly, M., 2007. 'Imagining Oceania: Indigenous and Foreign Representations of a Sea of Islands.' *Contemporary Pacific* 19: 508–545. doi.org/10.1353/cp.2007.0054

Nadasdy, P., 2007. 'Adaptive Co-management and the Gospel of Resilience.' In F. Berkes, D. Armitage and N. Doubleday (eds), *Adaptive Co-Management: Collaboration, Learning, and Multi-level Governance.* Seattle: University of Washington Press. doi.org/10.59962/9780774855457-014

Reid, W.V., F. Berkes, T.J. Wilbanks and D. Capistrano (eds), 2006. *Bridging Scales and Knowledge Systems: Concepts and Applications in Ecosystem Assessment.* Washington (DC): Island Press.

Steward, J.H., 1955. *Theory of Culture Change: The Methodology of Multilinear Evolution.* Urbana: University of Illinois Press.

Utz, R.J. (ed.), 2021. *Archipelagic Economies: Spatial Economic Development in the Pacific: Synthesis Report.* Washington (DC): World Bank. doi.org/10.1596/36749

Wesley-Smith, T., 2016. 'Rethinking Pacific Studies Twenty Years On.' *Contemporary Pacific* 28: 153–169. doi.org/10.1353/cp.2016.0003

www.ingramcontent.com/pod-product-compliance
Lightning Source LLC
Chambersburg PA
CBHW052001270326
41929CB00015B/2744